BRITISH RAIL

CO
STOCK

TWENTY-FOURTH EDITION
2000

The complete guide to all
Locomotive-Hauled Coaches which
operate on the Railtrack network

Peter Fox & Peter Hall

ISBN 1 902336 10 0

CONTENTS

UPDATES

An update to all the books in the *British Railways Pocket Book* series is published every month in the Platform 5 magazine, *Today's Railways*, which also contains news and rolling stock information on the railways of Britain , Ireland and Continental Europe. Rolling stock updates will also be found in other magazines specialising mainly in British matters, such as "Rail Express". For further details of *Today's Railways,* please see the advertisement inside the front cover of this book.

Information in this edition is intended to illustrate the actual situation on Britain's railways, rather than necessarily agree with TOPS and computer other records. Information is updated to 1 November 1999.

READERS' COMMENTS

With such a wealth of information as contained in this book, it is inevitable a few inaccuracies may be found. The author will be pleased to receive notification from readers of any such inaccuracies, and also notification of any additional information to supplement our records and thus enhance future editions.

Please send comments to: Peter Fox, Platform 5 Publishing Ltd., Wyvern House, Sark Road, Sheffield, S2 4HG, England.

Tel: 0114 255 2625 Fax: 0114 255 2471 e-mail: peter@platfive.freeserve.co.uk.

Both the author and the staff of Platform 5 regret they are unable to answer specific queries regarding locomotives and rolling stock other than through the 'Q & A' section in the Platform 5 magazine *Today's Railways*.

ORGANISATION & OPERATION OF BRITAIN'S RAILWAY SYSTEM

INFRASTRUCTURE & OPERATION

Britain's national railway infrastructure i.e. the track, signalling, stations and associated power supply equipment is owned by a public company – Railtrack PLC. Many stations and maintenance depots are leased to and operated by Train Operating Companies (TOCs), but some larger stations remain under Railtrack control. The only exception is the infrastructure on the Isle of Wight, which is nationally owned and is leased to the Island Line franchisee.

Trains are operated by TOCs over the Railtrack network, regulated by access agreements between the parties involved. In general, TOCs are responsible for the provision and maintenance of the locomotives, rolling stock and staff necessary for the direct operation of services, whilst Railtrack is responsible for the provision and maintenance of the infrastructure and also for staff needed to regulate the operation of services.

DOMESTIC PASSENGER TRAIN OPERATORS

The large majority of passenger trains are operated by the TOCs on fixed term franchises. Franchise expiry dates are shown in parentheses in the list of franchisees below:

Franchise	*Franchisee*	*Trading Name*
Anglia Railways	GB Railways Ltd. (until 4 April 2004)	Anglia Railways
Cardiff Railway	Prism Rail PLC (until 12 April 2004)	Cardiff Railways
Central Trains	National Express Group PLC (until 1 April 2004)	Central Trains
Chiltern Railways	M40 Trains Ltd. (until 20 July 2003)	Chiltern Railways
Cross Country	Virgin Rail Group Ltd. (until 4 January 2012)	Virgin Trains
Gatwick Express	National Express Group PLC (until 27 April 2011)	Gatwick Express
Great Eastern Railway	First Group PLC (until 4 April 2004)	First Great Eastern
Great Western Trains	First Group PLC (until 3 February 2006)	First Great Western
Inter City East Coast	GNER Holdings Ltd. (until 4 April 2004)	Great North Eastern Railway

Inter City West Coast	Virgin Rail Group Ltd. (until 8 March 2012)	Virgin Trains
Island Line	Stagecoach Holdings PLC (until 12 October 2001)	Island Line
LTS Rail	Prism Rail PLC (until 25 May 2011)	LTS Rail
Merseyrail Electrics	MTL Rail Ltd. (until 18 March 2004)	Merseyrail Electrics
Midland Main Line	National Express Group PLC (until 27 April 2006)	Midland Mainline
Network South Central	Connex Rail Ltd. (until 25 May 2003)	Connex South Central
North London Railways	National Express Group PLC (until 1 September 2004)	Silverlink Train Services
North West Regional Railways	First Group PLC (until 1 April 2004)	First North Western
Regional Railways North East	MTL Rail Ltd. (until 1 April 2004)	Northern Spirit
Scot Rail	National Express Group PLC (until 30 March 2004)	ScotRail
South Eastern	Connex Rail Ltd. (until 12 October 2011)	Connex South Eastern
South Wales & West	Prism Rail PLC (until 12 April 2004)	Wales & West Passenger Trains
South West	Stagecoach Holdings PLC (until 3 February 2003)	South West Trains
Thames	Victory Railways Holdings Ltd. (until 12 April 2004)	Thames Trains
Thameslink	GOVIA Ltd. (until 1 April 2004)	Thameslink Rail
West Anglia Great Northern	Prism Rail PLC (until 4 April 2004)	WAGN

The above companies may also operate other services under 'Open Access' arrangements.

The following operators run non-franchised services only:

Operator	*Trading Name*	*Route*
British Airports Authority	Heathrow Express	London Paddington–Heathrow Airport
West Coast Railway Co.	West Coast Railway	Fort William–Mallaig

INTERNATIONAL PASSENGER OPERATIONS

Eurostar (UK) operates international passenger-only services between the United Kingdom and continental Europe, jointly with French National Railways (SNCF) and Belgian National Railways (SNCB/NMBS). Eurostar (UK) is a subsidiary of London & Continental Railways, which is jointly owned by National Express Group PLC and the British Airports Authority.

In addition, a service for the conveyance of accompanied road vehicles through the Channel Tunnel is provided by the tunnel operating company, Eurotunnel.

FREIGHT TRAIN OPERATIONS

Freight train services are operated under 'Open Access' arrangements by English Welsh & Scottish Railway (EWS), Freightliner, Direct Rail Services and Mendip Rail.

INTRODUCTION

LAYOUT OF INFORMATION

Coaches are listed in numerical order of painted number in batches according to type.

Each coach entry is laid out as in the following example (former number column may be omitted where not applicable):

No.	*Prev. No.*	*Notes*	*Livery*	*Owner*	*Operation*	*Depot/Location*
2918	(40518)	*	**RP**	RT	*OR*	ZN

DETAILED INFORMATION & CODES

Under each type heading, the following details are shown:

- Diagram Code. This consists of the first three characters of the TOPS type code followed by two numbers which relate to the particular design of vehicle.
- 'Mark' of coach (see below).
- Descriptive text.
- Number of first class seats , standard class seats, lavatory compartments and wheelchair spaces shown as F/S nT nW respectively.
- Bogie type (see below).
- Additional features.
- ETH Index.

TOPS TYPE CODES

TOPS type codes are allocated to all coaching stock. For vehicles numbered in the passenger stock number series the code consists of:

(1) Two letters denoting the layout of the vehicle as follows:

AA Gangwayed Corridor
AB Gangwayed Corridor Brake
AC Gangwayed Open (2+2 seating)
AD Gangwayed Open (2+1 seating)
AE Gangwayed Open Brake
AF Gangwayed Driving Open Brake
AG Micro-Buffet
AH Brake Micro-Buffet
AI As 'AC' but with drop-head buckeye and gangway at one end only
AJ Restaurant Buffet with Kitchen

AK Kitchen Car
AL As 'AC' but with disabled person's toilet (Mark 4 only)
AN Miniature Buffet
AP Pullman First with Kitchen
AQ Pullman Parlour First
AR Pullman Brake First
AS Sleeping Car
AT Royal Train Coach
AU Sleeping Car with Pantry
AX Generator Van (1000 V d.c.)
AZ Special Saloon
GF DMU/EMU/Mark 4 Barrier Vehicle
AX Generator Van (415 V a.c three-phase)
NM Sandite Coach

(2) A digit denoting the class of passenger accommodation:

1 First
2 Standard (formerly second)
3 Composite (first & standard)
4 Unclassified
5 None

(3) A suffix relating to the build of coach.

1 Mark 1
Z Mark 2
A Mark 2A
B Mark 2B
C Mark 2C
D Mark 2D
E Mark 2E
F Mark 2F
G Mark 3 or 3A
H Mark 3B
J Mark 4

OPERATING CODES

Operating codes used by train company operating staff (and others) to denote vehicle types in general. These are shown in parentheses adjacent to TOPS type codes. Letters use are:

B Brake
C Composite
F First Class
K Side corridor with lavatory
O Open
S Standard Class (formerly second)

Various other letters are in use and the meaning of these can be ascertained by referring to the titles at the head of each type.

Readers should note the distinction between an SO (Open Standard) and a TSO (Tourist Open Standard) The former has 2 + 1 seating layout, whilst the latter has 2 + 2.

BOGIE TYPES

BR Mark 1 (BR1). Double bolster leaf spring bogie. Generally 90 m.p.h., but BR1 bogies may be permitted to run at 100 m.p.h. with special maintenance. Weight: 6.1 t.

BR Mark 2 (BR2). Single bolster leaf-spring bogie used on certain types of non-passenger stock and suburban stock (all now withdrawn). Weight: 5.3 t.

COMMONWEALTH (C). Heavy, cast steel coil spring bogie. 100 m.p.h. Weight: 6.75 t.

B4. Coil spring fabricated bogie. Generally 100 m.p.h., but B4 bogies may be permitted to run at 110 m.p.h. with special maintenance. Weight: 5.2 t.

B5. Heavy duty version of B4. 100 m.p.h. Weight: 5.3 t.

B5 (SR). A bogie originally used on Southern Region EMUs, similar in design to B5. Now also used on locomotive hauled coaches. 100 m.p.h.

BT10. A fabricated bogie designed for 125 m.p.h. Air suspension.

T4. A 125 m.p.h. bogie designed by BREL (now Adtranz).

BT41. Fitted to Mark 4 vehicles, designed by SIG in Switzerland. At present limited to 125 m.p.h., but designed for 140 m.p.h.

BRAKES

Air braking is now standard on British main line trains. Vehicles with other equipment are denoted:

v Vacuum braked.
x Dual braked (air and vacuum).

HEATING

Electric heating is now standard on British main-line trains. Certain coaches for use on charter services may in addition also have steam heating facilities, or be steam heated only.

PUBLIC ADDRESS

It is assumed all coaches are now fitted with public address equipment, although certain stored vehicles may not have this feature. In addition, it is assumed all vehicles with a conductor's compartment have public address transmission facilities, as have catering vehicles.

COOKING EQUIPMENT

It is assumed that Mark 1 catering vehicles have gas powered cooking equipment, whilst Mark 2, 3 and 4 catering vehicles have electric powered cooking equipment unless stated otherwise.

ADDITIONAL FEATURE CODES

d Secondary door locking.
dg Driver–Guard communication equipment.
f Facelifted or fluorescent lighting.
k Composition brake blocks (instead of cast iron).
n Day/night lighting.
p Public telephone.
pg Public address transmission and driver-guard communication.
pt Public address transmission facility.
q Catering staff to shore telephone.

w Wheelchair space.
z Disabled persons' toilet.

Standard class coaches with wheelchair space also have one tip-up seat per space.

NOTES ON ETH INDICES

The sum of ETH indices in a train must not be more than the ETS index of the locomotive. The normal voltage on British trains is 1000 V. Suffix 'X' denotes 600 amp wiring instead of 400 amp. Trains whose ETH index is higher than 66 must be formed completely of 600 amp wired stock. Class 55 locomotives cannot provide a consistent electric train supply for Mark 2E or 2D FO 3192/3202, FK 13585–13607 & BFK 17163–17172. Class 33 locomotives cannot provide a suitable electric train supply for Mark 2D, Mark 2E, Mark 2F, Mark 3, Mark 3A, Mark 3B or Mark 4 coaches.

BUILD DETAILS

Lot Numbers

Vehicles ordered under the auspices of BR were allocated a lot (batch) number when ordered and these are quoted in class headings and sub-headings.

Builders

These are shown in class headings, the following abbreviations being used:

ABB Derby	ABB, Derby Carriage Works (now Adtranz Derby).
ABB York	ABB, York.
Adtranz Derby	Adtranz, Derby.
Alstom Birmingham	Alstom, Saltley, Birmingham.
Alstom Eastleigh	Alstom, Eastleigh Works.
Ashford	BR, Ashford Works.
BRCW	Birmingham Railway Carriage & Wagon Co., Smethwick, Birmingham.
BREL Derby	BREL, Derby Carriage Works (later ABB Derby, now Adtranz Derby).
Charles Roberts	Charles Roberts and Co., Horbury, Wakefield (now Bombardier Prorail)
Cravens	Cravens, Sheffield.
Derby	BR, Derby Carriage Works (later BREL Derby, then ABB Derby, now Adtranz Derby).
Doncaster	BR, Doncaster Works (later BREL Doncaster, then BRML Doncaster, then ABB Doncaster, now Adtranz Doncaster)
Eastleigh	BR, Eastleigh Works (later BREL Eastleigh, then Wessex Traincare, now Alstom Eastleigh).
Glasgow	BR Springburn Works, Glasgow (now Railcare, Glasgow)
Gloucester	The Gloucester Railway Carriage & wagon Co.
Hunslet-Barclay	Hunslet Barclay, Kilmarnock Works
Metro-Cammell	Metropolitan-Cammell, Saltley, Birmingham (later GEC-A B'ham, now Alstom Birmingham).
Pressed Steel	Pressed Steel, Linwood.
Railcare Wolverton	Railcare, Wolverton.

Swindon	BR Swindon Works
Wessex Traincare	Wessex Traincare, Eastleigh Works (now Alstom Eastleigh).
Wolverton	BR Wolverton Works (later BREL Wolverton, now Railcare, Wolverton).
York	BR, York Carriage Works (later BREL York, then ABB York).

Information on sub-contracting works which built parts of vehicles e.g.the underframes etc. is not shown.

In addition to the above, certain vintage Pullman cars were built or rebuilt at the following works:

Metropolitan Carriage & Wagon Company, Birmingham (Now Alstom)
Midland Carriage & Wagon Company, Birmingham
Pullman Car Company, Preston Park, Brighton

Conversions have also been carried out at the Railway Technical centre, Derby, BR Salisbury Depot and Blakes Fabrications, Edinburgh.

Vehicle Numbers
Where a coach has been renumbered, the former number is shown in parentheses. If a coach has been renumbered more than once, the original number is shown first in parentheses, followed by the most recent previous number. Where the former number of a coach due to be converted or renumbered is known and the conversion and/or renumbering has not yet taken place, the coach is listed under both current number (with depot allocation) and under new number (without allocation).

Numbering Systems
Seven different numbering systems were in use on BR. These were the BR series, the four pre-nationalisation companies' series', the Pullman Car Company's series and the UIC (International Union of Railways) series. BR number series coaches, former Pullman Car Company series and UIC series coaches are listed separately. There is also a separate listing of 'Saloon' type vehicles which are registered to run on the Railtrack network. Please note the Mark 2 Pullman vehicles were ordered after the Pullman Car Company had been nationalised and are therefore numbered in the BR series.

THE DEVELOPMENT OF BR STANDARD COACHES

The standard BR coach built from 1951 to 1963 was the Mark 1. This type features a separate underframe and body. The underframe is normally 64 ft. 6 in. long, but certain vehicles were built on shorter (57 ft.) frames. Tungsten lighting was standard and until 1961, BR Mark 1 bogies were generally provided. In 1959 Lot No. 30525 (TSO) appeared with fluorescent lighting and melamine interior panels, and from 1961 onwards Commonwealth bogies were fitted in an attempt to improve the quality of ride which became very poor when the tyre profiles on the wheels of the BR1 bogies became worn. Later batches of TSO and BSO retained the features of Lot No. 30525, but compartment vehicles – whilst utilising melamine panelling in standard class – still retained tungsten lighting. Wooden interior finish was retained in first class vehicles where the only change was to fluorescent lighting in open vehicles (except Lot No. 30648, which had tungsten lighting). In later years many Mark 1 coaches had BR 1 bogies replaced by B4.

In 1964, a new prototype train was introduced. Known as 'XP64', it featured new seat designs, pressure heating & ventilation, aluminium compartment doors and corridor partitions, foot pedal operated toilets and B4 bogies. The vehicles were built on standard Mark 1 underframes. Folding exterior doors were fitted, but these proved troublesome and were later replaced with hinged doors. All XP64 coaches have been withdrawn, but some have been preserved.

The prototype Mark 2 vehicle (W 13252) was produced in 1963. This was an FK of semi-integral construction and had pressure heating & ventilation, tungsten lighting, and was mounted on B4 bogies. This vehicle has been preserved by the National Railway Museum. The production build was similar, but wider windows were used. The TSO and SO vehicles used a new seat design similar to that in the XP64 and fluorescent lighting was provided. Interior finish reverted to wood. Mark 2 vehicles were built from 1964–66.

The Mark 2A design, built 1967–68, incorporated the remainder of the features first used in the XP64 coaches, i.e. foot pedal operated toilets (except BSO), new first class seat design, aluminium compartment doors and partitions together with fluorescent lighting in first class compartments. Folding gangway doors (lime green coloured) were used instead of the traditional one-piece variety.

The following list summarises the changes made in the later Mark 2 variants:

Mark 2B: Wide wrap around doors at vehicle ends, no centre doors, slightly longer body. In standard class, one toilet at each end instead of two at one end as previously. Red folding gangway doors.

Mark 2C: Lowered ceiling with twin strips of fluorescent lighting and ducting for air conditioning, but air conditioning not fitted.

Mark 2D: Air conditioning. No opening top-lights in windows.

Mark 2E: Smaller toilets with luggage racks opposite. Fawn folding gangway doors.

Mark 2F: Plastic interior panels. Inter-City 70 type seats. Modified air conditioning system.

The Mark 3 design has BT10 bogies, is 75 ft. (23 m.) long and is of fully integral construction with Inter-City 70 type seats. Gangway doors were yellow (red in RFB) when new, although these are being changed on refurbishment. Loco-hauled coaches are classified Mark 3A, Mark 3 being reserved for HST trailers. A new batch of FO and BFO, classified Mark 3B, was built in 1985 with Advanced Passenger Train-style seating and revised lighting. The last vehicles in the Mark 3 series were the driving brake vans built for West Coast Main Line services.

The Mark 4 design was built by Metro-Cammell for use on the East Coast Main Line after electrification and features a body profile suitable for tilting trains, although tilt is not fitted, and is not intended to be. This design is suitable for 140 m.p.h. running, although is restricted to 125 m.p.h. pending installation of a more advanced signalling system on the route. The bogies for these coaches were built by SIG in Switzerland and are designated BT41. Power operated sliding plug exterior doors are standard.

1. BR NUMBER SERIES STOCK

AJ11 (RF) RESTAURANT FIRST

Dia. AJ106. Mark 1. 325 spent most of its life as a Royal Train vehicle and was numbered 2907 for a time. Built with Commonwealth bogies, but B5 bogies substituted on 325. 24/–. ETH 2.

Lot No. 30633 Swindon 1961. 42.5 t C, 41 t B5.

324	x	**CH**	NY	*ON*	NY
325		**PC**	VS	*ON*	SL

AP1Z (PK) PULLMAN FIRST WITH KITCHEN

Dia. AP101. Mark 2. Pressure Ventilated. 18/– 2T. B5 bogies. ETH 6.

Lot No. 30755 Derby 1966. 40 t.

Non-Standard Livery: Maroon & beige.

504	**0**	WC	*ON*	CS	THE WHITE ROSE
506	**0**	WC	*ON*	CS	THE RED ROSE

AQ1Z (PC) PULLMAN PARLOUR FIRST

Dia. AQ101. Mark 2. Pressure Ventilated. 36/– 2T. B4 bogies. ETH 5.

Lot No. 30754 Derby 1966. 35 t.

Non-Standard Livery: Maroon & beige.

546	**0**	WC	*ON*	CS	CITY OF MANCHESTER
548	**0**	WC	*ON*	CS	ELIZABETHAN
549	**0**	WC	*ON*	CS	PRINCE RUPERT
550	**0**	WC	*ON*	CS	GOLDEN ARROW
551	**0**	WC	*ON*	CS	CALEDONIAN
552	**0**	WC	*ON*	CS	SOUTHERN BELLE
553	**0**	WC	*ON*	CS	KING ARTHUR

AR1Z (PB) PULLMAN BRAKE FIRST

Dia. AR101. Mark 2. Pressure Ventilated. 30/– 2T. B4 bogies. ETH 4.

Lot No. 30753 Derby 1966. 35 t.

Non-Standard Livery: Maroon & beige.

586	**0**	WC	*ON*	CS	TALISMAN

AN21 (RG) GRIDDLE CAR

Dia. AN2??. Mark 1. Rebuilt from RF. –/30. B5 bogies. ETH 2.

This vehicle was numbered DB975878 for a time when in departmental service.

Lot No. 30013 Doncaster 1952. Rebuilt Wolverton 1965. 40 t.

1105	(302)	v	**G**	MH	*ON*	RL

AJ1F (RFB) BUFFET OPEN FIRST

Dia. AJ104. Mark 2F. Air conditioned. Converted 1988–9/91 at BREL, Derby from Mark 2F FOs. 1200/1/3/6/11/14–17/20/21/50/2/5/6/9 have Stones equipment, others have Temperature Ltd. 25/– 1T 1W (except 1217 and 1253 which are 26/– 1T). B4 bogies. p. q. d. ETH 6X.

1200/3/6/11/14/16/20/52/5/6. Lot No. 30845 Derby 1973. 33 t.
1201/4/5/7/8/10/12/13/15/17–9/21/50/1/4/7/9. Lot No. 30859 Derby 1973–4. 33 t.
1202/9/53/8. Lot No. 30873 Derby 1974–5. 33 t.

* Refurbished with new seat trim.
r Refurbished with new seat trim and new m.a. sets.

1200	(3287, 6459)	r	**V**	H	*VX*	MA
1201	(3361, 6445)	r	**V**	H	*VX*	MA
1202	(3436, 6456)	r	**V**	H	*VX*	MA
1203	(3291)	r		H	*VX*	MA
1204	(3401)	r	**V**	H	*VX*	MA
1205	(3329, 6438)	r	**V**	H	*VX*	MA
1206	(3319)	r	**V**	H	*VX*	MA
1207	(3328, 6422)	r	**V**	H	*VX*	MA
1208	(3393)	*	**V**	H	*VX*	MA
1209	(3437, 6457)	r	**V**	H	*VX*	MA
1210	(3405, 6462)	r	**V**	H	*VX*	MA
1211	(3305)	*		H	*VX*	MA
1212	(3427, 6453)	r	**V**	H	*VX*	MA
1213	(3419)	r	**V**	H	*VX*	MA
1214	(3317, 6433)	*		H	*VX*	MA
1215	(3377)	*		H	*VX*	MA
1216	(3302)	r	**V**	H	*VX*	MA
1217	(3357, 6444)			H	*SR*	IS
1218	(3332)			H	*AR*	NC
1219	(3418)			H	*AR*	NC
1220	(3315, 6432)	r	**V**	H	*VX*	MA
1221	(3371)	*		H	*VX*	MA
1250	(3372)	r	**V**	H	*VX*	MA
1251	(3383)	r	**V**	H	*VX*	MA
1252	(3280)	r	**V**	H	*VX*	MA
1253	(3432)	r	**V**	H	*VX*	MA
1254	(3391)	r	**V**	H	*VX*	MA
1255	(3284)	r	**V**	H	*VX*	MA
1256	(3296)	r		H	*VX*	MA
1258	(3322)	r	**V**	H	*VX*	MA
1259	(3439)	r	**V**	H	*VX*	MA
1260	(3378)	r	**V**	H	*VX*	MA

AK51 (RKB) KITCHEN BUFFET

Dia. AK502. Mark 1. No seats. B5 bogies. ETH 1.

Lot No. 30624 Cravens 1960–1. 41 t.

1566		**RB**	VS	*ON*	CP

AJ41 (RBR) RESTAURANT BUFFET

Dia. AJ403. Mark 1. Built with 23 loose chairs (Dia. AJ402). All remaining vehicles refurbished with 23 fixed polypropylene chairs and fluorescent lighting. ETH 2 (2X*).

r Further refurbished with 21 chairs, payphone, wheelchair space and carpets (Dia. AJ416).

1653–1699. Lot No. 30628 Pressed Steel 1960–61. Commonwealth bogies. 39 t.
1730. Lot No. 30512 BRCW 1960–61. B5 bogies. 37 t.

1653			CN		FK
1658		**BG**	RS	*ON*	BN
1659	x	**PC**	WT	*ON*	RL
1667	x		RS	*ON*	BN
1671	x*	**CC**	RS	*ON*	BN
1674			CN		BN
1679		**G**	RS	*ON*	BN
1680	x*	**WV**	RS	*ON*	BN
1683	r	**H**	H	*CA*	CF
1686	r		H		KN
1689	r		H		KN
1691	r	**G**	H		CP
1692	xr	**CH**	RV	*ON*	CP
1696		**G**	RS	*ON*	BN
1697	r		H		CP
1698		**WV**	RS	*ON*	BN
1699	r		H		CP
1730	x	**M**	SP	*ON*	BT

AN21 (RMB) MINIATURE BUFFET CAR

Dia. AN203. Mark 1. –/44 2T. These vehicles are basically an open standard with two full window spaces removed to accommodate a buffet counter, and four seats removed to allow for a stock cupboard. All remaining vehicles now have fluorescent lighting. All vehicles have Commonwealth bogies except 1850 (B5). ETH 3.

1813–1832. Lot No. 30520 Wolverton 1960. 38 t.
1840–1850. Lot No. 30507 Wolverton 1960. 37 t (1850 is 36 t).
1859–1863. Lot No. 30670 Wolverton 1961–2. 38 t.
1871–1882. Lot No. 30702 Wolverton 1962. 38 t.

1842/50/71 have been been refurbished and are fitted with a microwave oven and payphone. Dia. AN208.

1813	x	**CC**	RS	*ON*	BN
1832	x	**BG**	RS	*ON*	BN
1840	v	**G**	MH	*ON*	RL
1842	x		H	*AR*	NC
1850			H	*GW*	OO
1859	x	**M**	SP	*ON*	BT
1860	x	**M**	WC	*ON*	CS
1861	x	**M**	WC	*ON*	TM
1863	x	**CH**	RV	*ON*	CP
1871	x		H	*GW*	OO
1882	x	**M**	WC	*ON*	CS

AJ41 (RBR) RESTAURANT BUFFET

Dia. AJ414. Mark 1. This vehicle was built as an unclassified restaurant (RU). It was rebuilt with buffet counter and 23 fixed polypropylene chairs (RBS), then further refurbished by fitting fluorescent lighting and reclassified RBR. B4/B5 bogies. ETH 2X.

Lot No. 30575 Swindon 1960. 36.5 t.

1953 **RB** VS *ON* CP

AS41 FIRST CLASS SLEEPING CAR

Dia. AS101. Mark 1. Pressure Ventilated. 11 single-berth compartments plus an attendant's compartment. ETH 3 (3X*).

2013. Lot No. 30159 Wolverton 1958. B5 bogies. 39 t.
2127. Lot No. 30687 Wolverton 1961. Commonwealth bogies. 41 t.

2013 was numbered 2908 for a time when in use with the Royal Train.

2013 **M** FS SZ | 2127 * **M** GS CS

AU51 CHARTER TRAIN STAFF COACHES

Dia. AU501. Mark 1. Converted from BCKs in 1988. Commonwealth bogies. ETH 2.

Lot No. 30732 Derby 1964. 37 t.

2833 (21270) RS *ON* BN | 2834 (21267) **WV** RS *ON* BN

AT5G HM THE QUEEN'S SALOON

Dia. AT525. Mark 3. Converted from a FO built 1972. Consists of a lounge, bedroom and bathroom for HM The Queen, and a combined bedroom and bathroom for the Queen's dresser. One entrance vestibule has double doors. Air conditioned. BT10 bogies. ETH 9X.

Lot No. 30886 Wolverton 1977. 36 t.

2903 (11001) **RP** RT *OR* ZN

AT5G HRH THE DUKE OF EDINBURGH'S SALOON

Dia. AT526. Mark 3. Converted from a TSO built 1972. Consists of a combined lounge/dining room, a bedroom and a shower room for the Duke, a kitchen and a valet's bedroom and bathroom. Air conditioned. BT10 bogies. ETH 15X.

Lot No. 30887 Wolverton 1977. 36 t.

2904 (12001) **RP** RT *OR* ZN

AT5B ROYAL HOUSEHOLD COUCHETTES

Dia. AT527. Mark 2B. Converted from a BFK built 1969. Consists of luggage accommodation, guard's compartment, 350 kW diesel generator and staff sleeping accommodation. Pressure ventilated. B5 bogies. ETH 5X.

Lot No. 30888 Wolverton 1977. 46 t.

2905 (14105) **RP** RT *OR* ZN

Dia. AT528. Mark 2B. Converted from a BFK built 1969. Consists of luggage accommodation, guards compartment and staff accommodation. Pressure ventilated. B5 bogies. ETH 4X.

Lot No. 30889 Wolverton 1977. 35.5 t.

2906 (14112) **RP** RT *OR* ZN

AT5G ROYAL HOUSEHOLD SLEEPING CARS

Dia. AT531. Mark 3A. Built to similar specification as SLE 10646–732. 12 sleeping compartments for use of Royal Household with a fixed lower berth and a hinged upper berth. 2T plus shower room. Air conditioned. BT10 bogies. ETH 11X.

Lot No. 31002 Derby/Wolverton 1985. 42.5 t (44 t*).

2914 **RP** RT *OR* ZN
2915 * **RP** RT *OR* ZN

AT5G ROYAL KITCHEN/DINING CAR

Dia AT537. Mark 3. Converted from HST TRUK built 1976. Large kitchen retained, but dining area modified for Royal use seating up to 14 at central table(s). Air conditioned. BT10 bogies. ETH 13X.

Lot No. 31059 Wolverton 1988. 43 t.

2916 (40512) **RP** RT *OR* ZN

AT5G ROYAL HOUSEHOLD KITCHEN/DINING CAR

Dia AT539. Mark 3. Converted from HST TRUK built 1977. Large kitchen retained and dining area slightly modified with seating for 22 Royal Household members. Air conditioned. BT10 bogies. ETH 13X.

Lot No. 31084 Wolverton 1990. 43 t.

2917 (40514) **RP** RT *OR* ZN

AT5G ROYAL HOUSEHOLD CARS

Dia. AT538 (AT540*). Mark 3. Converted from HST TRUKs built 1976/7. Air conditioned. BT10 bogies. ETH 10X.

Lot Nos. 31083 (31085*) Wolverton 1989. 41.05 t.

2918	(40515)		**RP**	RT	*OR*	ZN
2919	(40518)	*	**RP**	RT	*OR*	ZN

AT5B ROYAL HOUSEHOLD COUCHETTES

Dia. AT536. Mark 2B. Converted from BFK built 1969. Consists of luggage accommodation, guard's compartment, workshop area, 350 kW diesel generator and staff sleeping accommodation. B5 bogies. ETH2X.

Lot No. 31044 Wolverton 1986. 48 t.

2920 (14109, 17109) **RP** RT *OR* ZN

Dia. AT541. Mark 2B. Converted from BFK built 1969. Consists of luggage accommodation, kitchen, brake control equipment and staff accommodation. B5 bogies. ETH7X.

Lot No. 31086 Wolverton 1990. 41.5 t.

2921 (14107, 17107) **RP** RT *OR* ZN

AT5G HRH THE PRINCE OF WALES'S SLEEPING CAR

Dia. AT534. Mark 3B. BT10 bogies. Air conditioned.ETH 7X.

Lot No. 31035 Derby/Wolverton 1987.

2922 **RP** RT *OR* ZN

AT5G HRH THE PRINCE OF WALES'S SALOON

Dia. AT535. Mark 3B. BT10 bogies. Air conditioned. ETH 6X.

Lot No. 31036 Derby/Wolverton 1987.

2923 **RP** RT *OR* ZN

AD11 (FO) OPEN FIRST

Dia. AD103. Mark 1. 42/– 2T. ETH 3. Many now fitted with table lamps.

3063–3069. Lot No. 30169 Doncaster 1955. B4 bogies. 33 t.
3096–3100. Lot No. 30576 BRCW 1959. B4 bogies. 33 t.

3064 and 3068 were numbered DB 975607 and DB 975606 for a time when in departmental service for British Rail. 3065 has BR Mark 1 bogies and weighs 34 t.

3063		**BG**	VS		SL
3064		**BG**	VS		SL
3065	v	**PC**	WT		CS
3066		**RB**	VS	*ON*	CP
3068		**RB**	VS	*ON*	CP
3069		**RB**	VS	*ON*	CP
3096	x	**M**	SP	*ON*	BT
3097		**WV**	RS	*ON*	BN
3098	x	**CH**	RV	*ON*	CP
3100	x	**CC**	RS	*ON*	BN

Later design with fluorescent lighting, aluminium window frames and Commonwealth bogies.

▲ **Mark 1 Stock.** Southern green liveried restaurant buffet No. 1696 is seen at Worcester Shrub Hill with a charter train from London King's Cross on 17th April 1999. This vehicle has Commonwealth bogies. **Stephen Widdowson**

▼ Open first No. 3105 'JULIA' in BR maroon livery stabled at Carlisle on 18th September 1999. This coach is owned by the West Coast Railway Company.
Kevin Conkey

Generator van No. 6313 now owned by VSOE is seen stabled at Worcester Shrub Hill on 25th October 1998. The vehicle is painted in Pullman umber and cream livery. **Stephen Widdowson**

▲ Carmine & cream liveried corridor brake composite No. 21245 is seen stabled at Worcester Shrub Hill on 26th June 1999. The vehicle is owned by Rail Charter Services. **Stephen Widdowson**

▼ **Mark 2A Stock.** Open standard (TSO) No. 5389 waits departure from Crewe as part of the stock of the 10.18 to Bangor on 14th August 1999. **Peter Fox**

▲ **Mark 2D Stock.** Open first No. 3181 'MONARCH' in Regency Railtours blue and cream livery at Reading on 17th June 1997. This coach is now owned by Venice Simplon-Orient Express. **Darren Ford**

▼ **Mark 2E Stock.** Virgin Trains liveried open standard No. 5899 at Carlisle on 21st June 1998 as part of the 17.00 Edinburgh–Birmingham New Street. **Kevin Conkey**

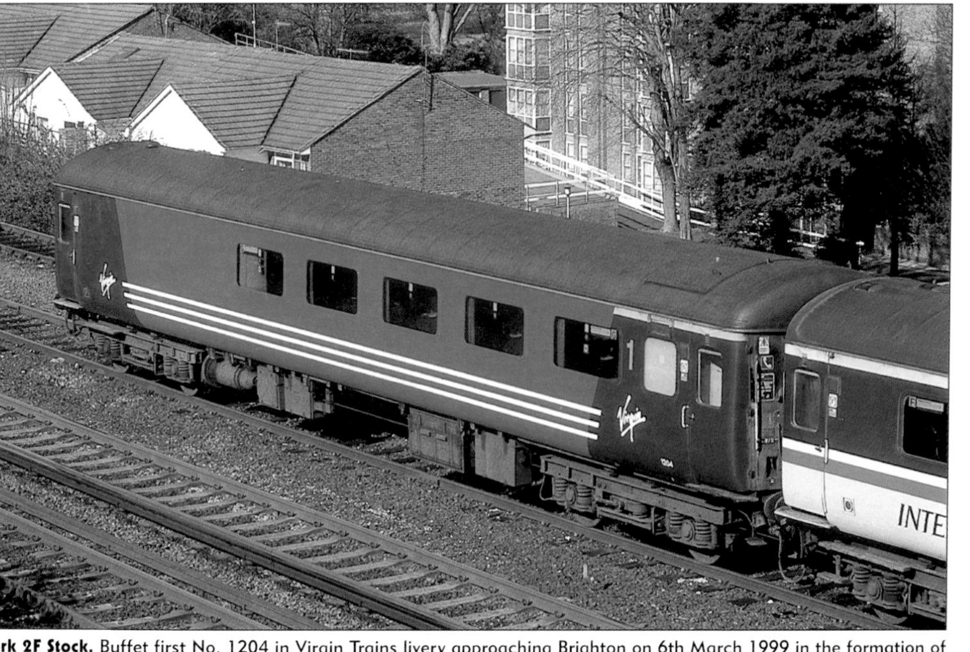

Mark 2F Stock. Buffet first No. 1204 in Virgin Trains livery approaching Brighton on 6th March 1999 in the formation of the 06.20 ex-Preston. **Chris Wilson**

Driving brake open standard No. 9709 pulls out of Colchester leading the 11.30 London Liverpool Street–Norwich service on 4th September 1999. **David Brown**

Mark 3 Stock. First Great Western (formerly Great Western Trains) has recently modified its livery with the addition of thin green stripes on the lower bodyside together with a broad gold band. Open standard No. 42029 was photographed at Dawlish on 4th July 1999.
Colin J. Marsden

▲ **Mark 3A Stock.** Restaurant buffet first No. 10223 in Anglia Railways livery at Colchester in the formation of the 11.30 London Liverpool Street–Norwich on 4th September 1999. Anglia has eight of these vehicles which operate in sets of Mark 2F stock. **David Brown**

▼ **Mark 3B Stock.** Virgin West Coast operate three BFO vehicles which were originally built for first class only Manchester Pullman workings. One of these, No. 17173 is seen at Carlisle on 25th July 1999. **Kevin Conkey**

3105–3128. Lot No. 30697 Swindon 1962–3. 36 t.
3130–3150. Lot No. 30717 Swindon 1963. 36 t.

3128/36/41/3/4/6/7/8 were renumbered 1058/60/3/5/6/8/9/70 when reclassified RUO, then 3600/5/8/9/2/6/4/10 when declassified, but have since regained their original numbers.

3105	x **M**	WC	*ON*	CS
3107	x **BG**	RS	*ON*	BN
3110	x **CC**	RS	*ON*	BN
3112	x **CH**	RV	*ON*	CP
3113	x **M**	WC	*ON*	CS
3114	**G**	RS	*ON*	BN
3115	x **BG**	RS	*ON*	BN
3117	x **M**	WC	*ON*	CS
3119	x **CC**	RS	*ON*	BN
3120	**WV**	RS	*ON*	BN
3121	**WV**	RS	*ON*	BN
3122	x **CH**	RV	*ON*	CP
3123	**G**	RS	*ON*	BN
3124	**G**	RS	*ON*	BN
3125	**RB**	VS	*ON*	CP
3127	**G**	RS	*ON*	BN
3128	x **M**	WC	*ON*	CS
3130	v **M**	WC	*ON*	CS
3131	x **CC**	RS	*ON*	BN
3132	x **CC**	RS	*ON*	BN
3133	x **CC**	RS	*ON*	BN
3136		RS	*ON*	BN
3140	x **CH**	RV	*ON*	CP
3141	**WV**	RS	*ON*	BN
3143		FS		SZ
3144	x **CC**	RS	*ON*	BN
3146	**WV**	RS	*ON*	BN
3147	**WV**	RS	*ON*	BN
3148	**BG**	RS	*ON*	BN
3149		RS	*ON*	BN
3150	**G**	RS	*ON*	BN

AD1D (FO) OPEN FIRST

Dia. AD105. Mark 2D. Air conditioned. 3172–88 have Stones equipment. 3192/3202 have Temperature Ltd and require at least 800 V train supply. 42/– 2T. B4 bogies. ETH 5.

Lot No. 30821 Derby 1971–2. 32.5 t.

3172		SO	*SO*	ZA
3174		VS	*ON*	CP
3178		VS		CP
3181	**RB**	VS	*ON*	CP
3182		VS		CP
3186		CN		DY
3187		E		KM
3188	**RB**	VS	*ON*	CP
3192		SO	*SO*	ZA
3202		E		KM

AD1E (FO) OPEN FIRST

Dia. AD106. Mark 2E. Air conditioned. Stones equipment. Require at least 800 V train supply. 42/– 2T (41/– 2T 1W w). B4 bogies. ETH 5.

* Seats removed to accommodate catering module. 40F 1T.
u Fitted with power supply for Mk. 1 RBR.

3255 was numbered 3525 for a time when fitted with a pantry.

Lot No. 30843 Derby 1972–3. 32.5 t.

3221	w	H		ZC
3223		CN		OM
3225		E		KN
3226		E		KN
3228	du	H	*GW*	OO
3229	d	H	*GW*	OO
3230		SO	*SO*	ZA
3231		RA		CP
3232	dw	H		ZC
3234	w	VS		CP
3235	u	H		PY
3237		CN		FK

3239			VS		CP
3240		CH	RV	*ON*	CP
3241	d	GW	H	*GW*	OO
3242	wu		H		PY
3244	d		H	*GW*	OO
3246	w		RA		CP
3247			VS		CP
3248			SO	*SO*	ZA
3251	*		CN		FK
3252	w		H		PY
3255	d	GW	H	*GW*	OO
3256	w		H		PY
3257	w		VS		CP
3258	n		E		KN
3261	dw		H	*GW*	OO
3267		CH	RV	*ON*	CP
3268			CN		KN
3269	d		H	*GW*	OO
3270			VS		CP
3272			VS		CP
3273		CH	RV	*ON*	CP
3275			VS		CP

AD1F (FO) OPEN FIRST

Dia. AD107. Mark 2F. Air conditioned. 3277–3318/58–81 have Stones equipment, others have Temperature Ltd. 42/– 2T. All now refurbished with power-operated vestibule doors, new panels and new seat trim. B4 bogies. d. ETH 5X.

3277–3318. Lot No. 30845 Derby 1973. 33 t.
3325–3428. Lot No. 30859 Derby 1973–4. 33 t.
3429–3438. Lot No. 30873 Derby 1974–5. 33 t.

r Further refurbished with table lamps and new burgundy seat trim.
u Fitted with power supply for Mk. 1 RBR.

3403 was numbered 6450 for a time when declassified.

3277			H	*AR*	NC
3278	r	**V**	H	*VW*	OY
3279	u	**AR**	H	*AR*	NC
3285	r	**V**	H	*VW*	OY
3290	r	**AR**	H	*AR*	NC
3292			H	*AR*	NC
3293			H	*GW*	OO
3295			H	*AR*	NC
3299	r	**V**	H	*VW*	OY
3300	r	**V**	H	*VW*	OY
3303		**AR**	H	*AR*	NC
3304	r	**V**	H	*VW*	OY
3309			H	*AR*	NC
3312			H	*GW*	OO
3313	r	**V**	H	*VW*	OY
3314	r	**V**	H	*VW*	OY
3318			H	*AR*	NC
3325	r	**V**	H	*VW*	OY
3326	r	**V**	H	*VW*	OY
3330	r	**V**	H	*VW*	OY
3331			H	*AR*	NC
3333	r	**V**	H	*VW*	OY
3334			H	*AR*	NC
3336	u	**AR**	H	*AR*	NC
3337	r	**V**	H	*VW*	OY
3338	ur	**AR**	H	*AR*	NC
3340	r	**V**	H	*VW*	OY
3344	r	**V**	H	*VW*	OY
3345	r	**V**	H	*VW*	OY
3348	r	**V**	H	*VW*	OY
3350	r	**V**	H	*VW*	OY
3351	r	**AR**	H	*AR*	NC
3352	r	**V**	H	*VW*	OY
3353	r	**V**	H	*VW*	OY
3354	r	**V**	H	*VW*	OY
3356	r	**V**	H	*VW*	OY
3358	r	**AR**	H	*AR*	NC
3359	r	**V**	H	*VW*	OY
3360	r	**V**	H	*VW*	OY
3362	r	**V**	H	*VW*	OY
3363	r	**V**	H	*VW*	OY
3364	r	**V**	H	*VW*	OY
3366	r	**V**	H	*VW*	OY
3368	r	**AR**	H	*AR*	NC
3369	r	**V**	H	*VW*	OY
3373			H	*AR*	NC
3374			H	*GW*	OO
3375			H	*AR*	NC

3379	u		H	*AR*	NC
3381			H	*AR*	NC
3384	r	**V**	H	*VW*	OY
3385	r	**V**	H	*VW*	OY
3386	r	**V**	H	*VW*	OY
3387	r	**V**	H	*VW*	OY
3388		**AR**	H	*AR*	NC
3389	r	**V**	H	*VW*	OY
3390	r	**V**	H	*VW*	OY
3392	r	**V**	H	*VW*	OY
3395	r	**V**	H	*VW*	OY
3397	r	**V**	H	*VW*	OY
3399	u	**AR**	H	*AR*	NC
3400	r	**AR**	H	*AR*	NC
3402	r	**V**	H	*VW*	OY
3403	r	**V**	H	*VW*	OY
3408	r	**V**	H	*VW*	OY
3411	r	**V**	H	*VW*	OY
3414		**AR**	H	*AR*	NC
3416			H	*AR*	NC
3417			H	*AR*	NC
3424		**AR**	H	*AR*	NC
3425	r	**V**	H	*VW*	OY
3426	r	**V**	H	*VW*	OY
3428	r	**V**	H	*VW*	OY
3429	r	**V**	H	*VW*	OY
3431	r	**V**	H	*VW*	OY
3433	r	**V**	H	*VW*	OY
3434	r	**V**	H	*VW*	OY
3438	r	**V**	H	*VW*	OY

AG1E (FO (T)) OPEN FIRST (PANTRY)

Dia. AG101. Mark 2E. Air conditioned. Converted from FO. Fitted with pantry, microwave oven and payphone for use on overnight services. 36/– 1T. B4 bogies. d. ETH 5X.

Lot No. 30843 Derby 1972–3. 32.5 t.

3520	(3253)	H	*GW*	LA
3521	(3271)	H	*GW*	LA
3522	(3236)	H	*GW*	LA
3523	(3238)	H	*SR*	IS
3524	(3254)	H	*SR*	IS

AC21 (TSO) OPEN STANDARD

Dia. AC204. Mark 1. These vehicles have 2+2 seating and are classified TSO ('Tourist second open'–a former LNER designation). –/64 2T. ETH 4.

3766. Lot No. 30079 York 1953. Commonwealth bogies (originally built with BR Mark 1 bogies). This coach has narrower seats than later vehicles. 36 t.
4198. Lot No. 30172 York 1956. BR Mark 1 bogies. 33 t.

3766	x	**M**	WC	*ON*	CS
4198	v	**CH**	NY	*ON*	NY

AD21 (SO) OPEN STANDARD

Dia. AD201. Mark 1. These vehicles have 2+1 seating and were often used as second class dining cars when new. –/48 2T. BR Mark 1 bogies. ETH 4.

4786. Lot No. 30376 York 1957. 33 t.
4817. Lot No. 30473 BRCW 1957–59. 33 t.

4786	v	**CH**	NY	*ON*	NY
4817	v	**CH**	NY	*ON*	NY

AC21 (TSO) OPEN STANDARD

Dia. AC201. Mark 1. These vehicles are a development of Dia. AC204 with fluorescent lighting and modified design of seat headrest. Built with BR Mark 1 bogies. –/64 2T. ETH 4.

4831–4836. Lot No. 30506 Wolverton 1959. Commonwealth bogies. 33 t.
4849–4880. Lot No. 30525 Wolverton 1959–60. B4 bogies. 33 t.

4831	x	**M**	SP	*ON*	BT
4832	x	**M**	SP	*ON*	BT
4836	x	**M**	SP	*ON*	BT
4849		**RR**	H	*NW*	CP
4854		**RR**	H	*NW*	CP
4856	X	**M**	SP	*ON*	BT
4866		**RR**	H	*NW*	CP
4869	x		CN		FK
4873		**RR**	H	*NW*	CP
4875		**RR**	H	*NW*	CP
4876		**RR**	H	*NW*	CP
4880		**RR**	H	*NW*	CP

Lot No. 30646 Wolverton 1961. Built with Commonwealth bogies, but BR Mark 1 bogies substituted by the SR on 4902/5/9/10/12/15/16. All now re-rebogied. 34 t B4, 36 t C.

4902	x B4	**CH**	RV	*ON*	CP
4905	x C	**M**	WC	*ON*	CS
4909	x B4		CN		FK
4912	x C	**M**	WC	*ON*	CS
4915	x B4	**CC**	RS	*ON*	BN
4916	x B4	**CC**	RS	*ON*	BN
4917	x C	**RR**	H		CP

Lot No. 30690 Wolverton 1961–2. Commonwealth bogies and aluminium window frames. 37 t.

4925		**G**	RS	*ON*	BN
4927	x	**CH**	RV	*ON*	CP
4931	v	**M**	WC	*ON*	CS
4938		**BG**	RS	*ON*	BN
4939			RS	*ON*	BN
4940	x	**M**	WC	*ON*	CS
4946	x	**CC**	RS	*ON*	BN
4949		**BG**	RS	*ON*	BN
4951	x	**M**	WC	*ON*	CS
4954	v	**M**	WC	*ON*	CS
4956		**BG**	RS	*ON*	BN
4958	v	**M**	WC	*ON*	CS
4959			RS	*ON*	BN
4960	x	**M**	WC	*ON*	CS
4963	x	**CH**	RV	*ON*	CP
4973	x	**M**	WC	*ON*	CS
4977			RS	*ON*	BN
4984	v	**M**	WC	*ON*	CS
4986		**G**	RS	*ON*	BN
4991		**BG**	RS	*ON*	BN
4993			CN		FK
4994	x	**M**	WC	*ON*	CS
4996	x	**CC**	RS	*ON*	BN
4998			RS	*ON*	CP
4999		**BG**	RS	*ON*	BN
5002		**WR**	RS	*ON*	CP
5005		**BG**	RS	*ON*	BN
5007		**G**	RS	*ON*	BN
5008	x	**CC**	RS	*ON*	BN
5009	x	**CH**	RV	*ON*	CP
5010			RV		CP
5023		**G**	RS	*ON*	BN
5025	x	**CH**	RV	*ON*	CP
5027		**G**	RS	*ON*	BN
5028	x	**M**	SP	*ON*	BT
5029	x	**CH**	RV	*ON*	CP
5030	x	**CH**	RV	*ON*	CP
5032	x	**M**	WC	*ON*	CS
5033	x	**M**	WC	*ON*	CS
5035	x	**M**	WC	*ON*	CS
5037		**G**	RS	*ON*	BN
5040	x	**CH**	RV	*ON*	CP
5042	x		CN		FK
5044	x	**M**	WC	*ON*	CS

AC2Z (TSO) OPEN STANDARD

Dia. AC205. Mark 2. Pressure ventilated. –/64 2T. B4 bogies. ETH 4.

Lot No. 30751 Derby 1965–7. 32 t.

5125	v	**G**	MH	*ON*	RL
5132	v	**LN**	H		LT

5135	v	**RR**	H		LT
5148	v	**RR**	BM		TM
5154	v	**LN**	H		LT
5156	v	**RR**	H		LT
5157	v	**RR**	BM		TM
5158	v	**RR**	H		LT
5161	v	**RR**	H		LT
5163	v	**RR**	H		LT
5166	v	**LN**	H		LT
5167	v	**RR**	H		LT
5171	v	**G**	MH	*ON*	RL
5174	v	**RR**	H		LT
5177	v	**RR**	H		LT
5179	v	**RR**	BR		TM
5180	v	**RR**	H		LT
5183	v	**RR**	H		LT
5186	v	**RR**	BM		TM
5191	v	**LN**	BM		TM
5193	v	**LN**	BM		TM
5194	v	**RR**	BM		TM
5198	v	**CH**	BM	*ON*	TM
5200	v	**G**	MH	*ON*	RL
5207	v	**RR**	H		LT
5209	v	**RR**	H		LT
5212	v	**LN**	**BM**		TM
5213	v	**RR**	H		LT
5216	v	**G**	MH	*ON*	RL
5221	v	**RR**	BR		TM
5222	v	**G**	MH	*ON*	RL
5225	v	**RR**	H		LT
5226	v	**RR**	H		LT

AD2Z (SO) OPEN STANDARD

Dia. AD203. Mark 2. Pressure ventilated. –/48 2T. B4 bogies. ETH 4.

Lot No. 30752 Derby 1966. 32 t.

5237	v	**G**	MH	*ON*	RL
5249	v	**G**	MH	*ON*	RL
5254		**BG**	H		DY

AC2A (TSO) OPEN STANDARD

Dia. AC206. Mark 2A. Pressure ventilated. –/64 2T (–/62 2T w). B4 bogies. ETH 4.

5265–5345. Lot No. 30776 Derby 1967–8. 32 t.
5350–5433. Lot No. 30787 Derby 1968. 32 t.

5265		**RR**	H		KN
5266		**RR**	H		Crewe S. Yd
5267		**RR**	H		KN
5271		**RR**	H		KN
5272		**RR**	H		CP
5275		**H**	H	*CA*	CF
5276		**RR**	H		CP
5277		**BG**	H		KN
5278		**RR**	H	*NW*	CP
5282		**RR**	H		KN
5290		**NB**	H		KN
5292		**RR**	H		CP
5293		**NB**	H		KN
5299		**M**	WC	*ON*	CS
5304		**RR**	H		Crewe S. Yd
5307		**H**	H	*CA*	CF
5309		**RR**	H	*NW*	CP
5322		**RR**	H		CP
5331		**RR**	H	*NW*	CP
5335		**RR**	H	*NW*	CP
5341		**RR**	H		CP
5345		**RR**	H	*NW*	CP
5350		**H**	H	*CA*	CF
5353		**RR**	H		KN
5354		**RR**	H		PY
5364		**H**	H	*CA*	CF
5365		**H**	H	*CA*	CF
5366		**RR**	H		KN
5373		**H**	H	*CA*	CF
5376		**H**	H	*CA*	CF
5378		**H**	H	*CA*	CF
5379		**RR**	H		KN
5381	w	**RR**	H	*NW*	CP
5384		**N**	H		Crewe S. Yd
5386	w	**RR**	H	*NW*	CP
5389	w	**RR**	H	*NW*	CP
5396		**RR**	H		KN
5410		**N**	H		KN
5412	w	**RR**	H	*NW*	CP
5419	w	**RR**	H	*NW*	CP

5420	w	**RR**	H	*NW*	CP	5433	w	**RR**	H		CP

AC2B (TSO) OPEN STANDARD

Dia. AC207. Mark 2B. Pressure ventilated. –/62 2T. B4 bogies. ETH 4.

Lot No. 30791 Derby 1969. 32 t.

Non-Standard Livery: 5453, 5478 and 5491 are royal blue with white lining.

5439		**N**	H		KN	5464		**N**	RV		CP
5443		**N**	H		KN	5471		**N**	H		KN
5446		**N**	H		KN	5472		**N**	H		KN
5447		**N**	H		PY	5475		**N**	H		KN
5449		**N**	RV		CP	5478	d	**0**	WC	*WW*	CF
5450		**N**	H		KN	5480		**N**	H		KN
5453	d	**0**	WC	*WW*	CF	5487	d	**M**	WC	*WW*	CF
5454		**N**	H		KN	5491	d	**0**	WC	*WW*	CF
5462		**N**	RV		CP	5494		**N**	RV		CP
5463	d	**M**	WC	*WW*	CF						

AC2C (TSO) OPEN STANDARD

Dia. AC208. Mark 2C. Pressure ventilated. –/62 2T. B4 bogies. ETH 4.

Lot No. 30795 Derby 1969–70. 32 t.

5554		**RR**	H		CW	5600		**M**	WC	*ON*	CS
5569	d	**M**	WC	*WW*	CF	5614		**RR**	H		Crewe S. Yd

AC2D (TSO) OPEN STANDARD

Dia. AC209. Mark 2D. Air conditioned. Stones equipment. –/62 2T. B4 bogies. ETH 5.

Non-Standard Livery: 5630, 5647 & 5739 are **WV** without lining.

r Refurbished with new seats and centre luggage stack. –/58 2T.

Lot No. 30822 Derby 1971. 33 t.

5616			CN		FK	5647		**0**	RV	*ON*	CP
5618			H		PY	5650			H		LT
5620			H		LT	5657	dr	**GW**	H	*GW*	OO
5623			H		LT	5661			H		KN
5628			H		ZC	5662			H		ZN
5629			H		LT	5663			H		KN
5630		**0**	RV	*SL*	BY	5665			H		ZC
5631	d		H		ZC	5669	d		H		ZC
5632	d		H		ZC	5674			H		KN
5634			H		ZC	5676			H		ZC
5636	dr	**GW**	H	*GW*	OO	5679	d		H		ZC
5640			H		LT	5686			H		ZC
5642			WC		CS	5687			H		KN
5645			WC		CS	5690			H		LT

5692			H		ZC
5694			H		KN
5699			H		KN
5700	d		H		ZC
5701			H		KN
5704		**M**	WC	*ON*	CS
5709		**BG**	WC		CS
5710	d		H		ZC
5711			H		LT
5712			WC		CS
5714		**M**	WC	*ON*	CS
5715			H		LT
5716			H		KN
5718			H		KN
5722			E		KM
5724			H		LT
5726			H		PY
5727		**M**	WC	*ON*	CS
5728			H		PY
5731			H		KN
5732		**CH**	RV	*SL*	BY
5735			H		ZC
5737	dr	**GW**	H	*GW*	OO
5738			H		KN
5739		**0**	RV		CP
5740	d		H		ZC

AC2E (TSO) OPEN STANDARD

Dia. AC210. Mark 2E. Air conditioned. Stones equipment. Require at least 800 V train supply. –/64 2T (w –/62 2T 1W). B4 bogies. d (except 5756 and 5879). ETH 5.

5744–5801. Lot No. 30837 Derby 1972. 33.5 t.
5810–5906. Lot No. 30844 Derby 1972–3. 33.5 t.

r Refurbished with new green seat trim.
s Refurbished with new green seat trim, modified design of seat headrest and centre luggage stack. –/60 2T (w –/58 2T 1W).

5744			H	*GW*	OO
5745	s	**V**	H	*VX*	MA
5746	r	**V**	H	*VX*	MA
5748	r pt		H	*VX*	MA
5750	s	**V**	H	*VX*	MA
5752	rw		H	*VX*	MA
5754	sw	**V**	H	*VX*	MA
5756		**M**	WC	*ON*	CS
5769	r		H	*VX*	MA
5773	s pt	**V**	H	*VX*	MA
5775	s	**V**	H	*VX*	MA
5776	r		H	*VX*	MA
5778	w		H	*GW*	OO
5779	r		H	*VX*	MA
5780	w		H	*GW*	OO
5781	w		H	*GW*	OO
5784	r	**V**	H	*VX*	MA
5787	s	**V**	H	*VX*	MA
5788	r		H	*VX*	MA
5789	r pt		H	*VX*	MA
5791	rw		H	*VX*	MA
5792	r		H	*VX*	MA
5793	swpt	**V**	H	*VX*	MA
5794	rw		H	*VX*	MA
5796	rw		H	*VX*	MA
5797	r		H	*VX*	MA
5800			H	*GW*	OO
5801	r	**V**	H	*VX*	MA
5810	s	**V**	H	*VX*	MA
5812	rw		H	*VX*	MA
5814	r		H	*VX*	MA
5815	sw	**V**	H	*VX*	MA
5816	r pt		H	*VX*	MA
5821	r pt	**V**	H	*VX*	MA
5822	swpt	**V**	H	*VX*	MA
5824	rw		H	*VX*	MA
5827	r		H	*VX*	MA
5828	sw	**V**	H	*VX*	MA
5831			H	*GW*	OO
5836			H	*GW*	OO
5843	rw		H	*VX*	MA
5845	s	**V**	H	*VX*	MA
5847	rw	**V**	H	*VX*	MA
5852			H	*GW*	OO
5853			H	*AR*	NC
5854	r		H	*VX*	MA
5859	s	**V**	H	*VX*	MA
5863			H	*GW*	OO
5866	r pt		H	*VX*	MA
5868	s pt	**V**	H	*VX*	MA
5869			H	*AR*	NC
5874	w		H	*AR*	NC

5876	s pt	**V**	H	*VX*	MA
5879			E		OM
5881	sw	**V**	H	*VX*	MA
5886	s	**V**	H	*VX*	MA
5887	rw		H	*VX*	MA
5888	rw		H	*VX*	MA
5889	s	**V**	H	*VX*	MA
5893	s	**V**	H	*VX*	MA
5897	r		H	*VX*	MA
5899	s	**V**	H	*VX*	MA
5900	sw	**V**	H	*VX*	MA
5901	s	**V**	H	*VX*	MA
5902	s	**V**	H	*VX*	MA
5903	s	**V**	H	*VX*	MA
5905	s	**V**	H	*VX*	MA
5906	swpt		H	*VX*	MA

AC2F (TSO) OPEN STANDARD

Dia. AC211. Mark 2F. Air conditioned. Temperature Ltd. equipment. Inter-City 70 seats. All were refurbished in the 1980s with power-operated vestibule doors, new panels and new seat trim. –/64 2T. (w –/62 2T 1W) B4 bogies. d. ETH 5X.

5908–5958. Lot No. 30846 Derby 1973. 33 t.
5959–6170. Lot No. 30860 Derby 1973–4. 33 t.
6171–6184. Lot No. 30874 Derby 1974–5. 33 t.

* Early Mark 2 style seats.

These vehicles are now undergoing a second refurbishment with carpets, new motor-alternator sets and new seat trim.

r Standard refurbished vehicles.

Cross-Country vehicles:

s Fitted with centre luggage stack. –/60 2T.
t Fitted with centre luggage stack and wheelchair space. –/58 2T 1W.

West Coast vehicles:
u Refurbished with two wheelchair spaces. –/60 2T 2W.

5908	r	**V**	H	*VW*	OY
5910	u	**V**	H	*VW*	OY
5911	s	**V**	H	*VX*	MA
5912	s	**V**	H	*VX*	MA
5913	s		H	*VX*	MA
5914	u	**V**	H	*VW*	OY
5915	r	**V**	H	*VW*	OY
5916	t		H	*VX*	MA
5917	s	**V**	H	*VX*	MA
5918	t	**V**	H	*VX*	MA
5919	s pt	**V**	H	*VX*	MA
5920	r	**V**	H	*VW*	OY
5921			H	*AR*	NC
5922		**AR**	H	*AR*	NC
5924		**AR**	H	*AR*	NC
5925	s pt		H	*VX*	MA
5926			H	*AR*	NC
5927		**AR**	H	*AR*	NC
5928		**AR**	H	*AR*	NC
5929			H	*AR*	NC
5930	t	**V**	H	*VX*	MA
5931	rw	**V**	H	*VW*	OY
5932	r	**V**	H	*VW*	OY
5933	r	**V**	H	*VW*	OY
5934	r	**V**	H	*VW*	OY
5935		**AR**	H	*AR*	NC
5936		**AR**	H	*AR*	NC
5937	r	**V**	H	*VW*	OY
5939	r	**V**	H	*VW*	OY
5940	u	**V**	H	*VW*	OY
5941	r	**V**	H	*VW*	OY
5943	rw	**V**	H	*VW*	OY
5944	w	**AR**	H	*AR*	NC
5945	r	**V**	H	*VW*	OY
5946	r	**V**	H	*VW*	OY
5947	s pt	**V**	H	*VX*	MA
5948	u	**V**	H	*VW*	OY
5949	u	**V**	H	*VW*	OY

5950		H	*AR*	NC
5951	r	**V** H	*VX*	MA
5952	r	**V** H	*VW*	OY
5953	r	**V** H	*VW*	OY
5954		**AR** H	*AR*	NC
5955	r	**V** H	*VW*	OY
5956		H	*AR*	NC
5957	r	**V** H	*VW*	OY
5958	s	H	*VX*	MA
5959	n	**AR** H	*AR*	NC
5960	s	**V** H	*VX*	MA
5961	s pt	**V** H	*VX*	MA
5962	s pt	**V** H	*VX*	MA
5963	r	**V** H	*VW*	OY
5964		H	*AR*	NC
5965	t	H	*VX*	MA
5966		**AR** H	*AR*	NC
5967	t	**V** H	*VX*	MA
5968		**AR** H	*AR*	NC
5969	u	**V** H	*VW*	OY
5971	s	H	*VX*	MA
5973		**AR** H	*AR*	NC
5975	s	**V** H	*VX*	MA
5976	t	**V** H	*VX*	MA
5977	r	**V** H	*VW*	OY
5978	r	**V** H	*VW*	OY
5980	r	**V** H	*VW*	OY
5981	s	H	*VX*	MA
5983	s	**V** H	*VX*	MA
5984	r	**V** H	*VW*	OY
5985		H	*AR*	NC
5986	r	**V** H	*VW*	OY
5987	r	**V** H	*VW*	OY
5988	r	**V** H	*VW*	OY
5989	t	**V** H	*VX*	MA
5991	s	**V** H	*VX*	MA
5993	w*	**AR** H	*AR*	NC
5994	r	**V** H	*VX*	MA
5995	s	H	*VX*	MA
5996	s pt	**V** H	*VX*	MA
5997	r	**V** H	*VW*	OY
5998		H	*AR*	NC
5999	s	**V** H	*VX*	MA
6000	t	**V** H	*VX*	MA
6001	u	**V** H	*VW*	OY
6002	r	**V** H	*VW*	OY
6005	r	**V** H	*VX*	MA
6006		**AR** H	*AR*	NC
6008	s	**V** H	*VX*	MA
6009	r	**V** H	*VW*	OY
6010	s	**V** H	*VX*	MA
6011	s	**V** H	*VX*	MA
6012	r	**V** H	*VW*	OY
6013	s	H	*VX*	MA
6014	s pt	H	*VX*	MA
6015	t	**V** H	*VX*	MA
6016	r	**V** H	*VW*	OY
6018	t	**V** H	*VX*	MA
6021	r	**V** H	*VW*	OY
6022	s	**V** H	*VX*	MA
6024	s	**V** H	*VX*	MA
6025	t	**V** H	*VX*	MA
6026	s	**V** H	*VX*	MA
6027	u	**V** H	*VW*	OY
6028		**AR** H	*AR*	NC
6029	r	**V** H	*VW*	OY
6030	t	**V** H	*VX*	MA
6031	r	**V** H	*VW*	OY
6034		**AR** H	*AR*	NC
6035	t	H	*VX*	MA
6036	*	**AR** H	*AR*	NC
6037		**AR** H	*AR*	NC
6038	s	**V** H	*VX*	MA
6041	s	**V** H	*VX*	MA
6042		H	*AR*	NC
6043	r	**V** H	*VW*	OY
6045	rw	**V** H	*VW*	OY
6046	s	**V** H	*VX*	MA
6047	rn*	**V** H	*VW*	OY
6049	r	**V** H	*VW*	OY
6050	s	H	*VX*	MA
6051	r	**V** H	*VW*	OY
6052	tw	H	*VX*	MA
6053	*	H	*AR*	NC
6054	r	**V** H	*VW*	OY
6055	r	**V** H	*VW*	OY
6056	r	**V** H	*VW*	OY
6057	r	**V** H	*VW*	OY
6059	s	**V** H	*VX*	MA
6060	u	**V** H	*VW*	OY
6061	s pt	**V** H	*VX*	MA
6062	r	**V** H	*VW*	OY
6063	rw	**V** H	*VW*	OY
6064	s	**V** H	*VX*	MA
6065	r	**V** H	*VW*	OY
6066	s	H	*VX*	MA
6067	s pt	**V** H	*VX*	MA
6073	s	**V** H	*VX*	MA
6100	r*	**V** H	*VW*	OY
6101	r	**V** H	*VW*	OY
6102	r	**V** H	*VW*	OY
6103		H	*AR*	NC

6104	r	**V**	H	*VW*	OY
6105	t pt	**V**	H	*VX*	MA
6106	r	**V**	H	*VW*	OY
6107	r	**V**	H	*VW*	OY
6110	w		H	*AR*	NC
6111	r	**V**	H	*VW*	OY
6112	s pt	**V**	H	*VX*	MA
6113	r	**V**	H	*VW*	OY
6115	s		H	*VX*	MA
6116	r	**V**	H	*VW*	OY
6117	t	**V**	H	*VX*	MA
6119	s	**V**	H	*VX*	MA
6120	s	**V**	H	*VX*	MA
6121	r	**V**	H	*VW*	OY
6122	s	**V**	H	*VX*	MA
6123			H	*AR*	NC
6124	s pt		H	*VX*	MA
6134	r	**V**	H	*VW*	OY
6135	s		H	*VX*	MA
6136	r	**V**	H	*VW*	OY
6137	s pt	**V**	H	*VX*	MA
6138	r	**V**	H	*VW*	OY
6139	n*		H	*AR*	NC
6141	u	**V**	H	*VW*	OY
6142	r*	**V**	H	*VW*	OY
6144	r*	**V**	H	*VW*	OY
6145	s pt	**V**	H	*VX*	MA
6146	*		H	*AR*	NC
6147	s	**V**	H	*VW*	OY
6148	s		H	*VX*	MA
6149	u	**V**	H	*VW*	OY
6150	s		H	*VX*	MA
6151	r*	**V**	H	*VW*	OY
6152		**AR**	H	*AR*	NC
6153	r	**V**	H	*VW*	OY
6154	r pt		H	*VX*	MA
6155	*	**AR**	H	*AR*	NC
6157	r	**V**	H	*VX*	MA
6158	r	**V**	H	*VW*	OY
6159	s pt	**V**	H	*VX*	MA
6160	*		H	*AR*	NC
6161	r*	**V**	H	*VW*	OY
6162	s pt	**V**	H	*VX*	MA
6163	r	**V**	H	*VW*	OY
6164	r	**V**	H	*VW*	OY
6165	r	**V**	H	*VW*	OY
6166			H	*AR*	NC
6167		**AR**	H	*AR*	NC
6168	s		H	*VX*	MA
6170	s	**V**	H	*VX*	MA
6171	r	**V**	H	*VW*	OY
6172	s	**V**	H	*VX*	MA
6173	s	**V**	H	*VX*	MA
6174			H	*AR*	NC
6175	r	**V**	H	*VW*	OY
6176	t	**V**	H	*VX*	MA
6177	s	**V**	H	*VX*	MA
6178	s		H	*VX*	MA
6179	r	**V**	H	*VW*	OY
6180	rw	**V**	H	*VW*	OY
6181	rwn	**V**	H	*VW*	OY
6182	s	**V**	H	*VX*	MA
6183	s	**V**	H	*VX*	MA
6184	s	**V**	H	*VX*	MA

AC2D (TSO) OPEN STANDARD

Dia. AC217. Mark 2D. Air conditioned (Stones). Rebuilt from FO with new style 2+2 seats. –/58 2T. (–/58 1T*). B4 bogies. d. ETH 5X.

Lot No. 30821 Derby 1971–2. 33.5 t.

* One toilet converted to store room for use of attendant on sleeping car services.

6200	(3198)		H	*GW*	OO
6202	(3191)	*	H		CP
6203	(3180)		H	*GW*	OO
6206	(3183)		H	*GW*	LA
6207	(3204)		H	*GW*	OO
6212	(3176)		H	*GW*	OO
6213	(3208)		H	*GW*	LA
6219	(3213)		H	*GW*	OO
6221	(3173)		H		CP
6226	(3203)		H	*GW*	LA

GX51 GENERATOR VAN

Dia. GX501. Mark 1. Renumbered 1989 from BR departmental series. Converted from NDA in 1973 to three-phase supply generator van for use with HST trailers. Currently used to test overhauled HST trailers. B4 bogies.

Lot No. 30400 Pressed Steel 1958.

6310	(81448, 975325)	**P**	P	*P*	ZG

AX51 GENERATOR VAN

Dia. AX501. Mark 1. Converted from NDA in 1992 to generator vans for use on Anglo-Scottish sleeping car services. Now normally used on trains hauled by steam locomotives. B4 bogies. ETH75.

6311. Lot No. 30162 Pressed Steel 1958. 37.25 t.
6312. Lot No. 30224 Cravens 1956. 37.25 t.
6313. Lot No. 30484 Pressed Steel 1958. 37.25 t.

Non-Standard Livery: 6311 is purple.

6313 is leased to the Venice Simplon Orient Express.

6311	(80903, 92911)	**B**	RS	*ON*	BN
6312	(81023, 92925)		FS		SZ
6313	(81553, 92167)	**PC**	P	*ON*	SL

GS5 (HSBV) HST BARRIER VEHICLE

Various diagrams. Renumbered from departmental stock, or converted from various types. B4 bogies (Commonwealth bogies *).

6330. Mark 2A. Lot No. 30786 Derby 1968. 32 t.
6334. Mark 1. Lot No. 30400 Pressed Steel 1957–8. 31.5 t.
6336/8/44. Mark 1. Lot No. 30715 Gloucester 1962. 31 t.
6340. Mark 1. Lot No. 30669 Swindon 1962. 36 t.
6346. Mark 2A. Lot No. 30777 Derby 1967. 31.5 t.
6347. Mark 2A. Lot No. 30787 Derby 1968. 31.5 t.
6348. Mark 1. Lot No. 30163 Pressed Steel 1957. 31.5 t.

6330	(14084, 975629)		**G**	A	*A*	LA
6334	(81478, 92128)		**P**	P	*P*	NL
6336	(81591, 92185)			A	*A*	LA
6338	(81581, 92180)		**G**	A	*A*	LA
6340	(21251, 975678)	*	**G**	A	*A*	LA
6344	(81263, 92080)		**B**	A	*A*	EC
6346	(9422)		**B**	A	*A*	EC
6347	(5395)			A	*A*	LA
6348	(81233, 92963)		**G**	A	*A*	LA

GF5 (MFBV) MARK 4 BARRIER VEHICLE

Various diagrams. Renumbered from departmental stock, or converted from FK, BSO or BG. B4 bogies.

6351. Mark 1. Lot No. 30091 Doncaster 1954. 33 t.
6352/3. Mark 2A. Lot No. 30774 Derby 1968. 33 t.
6354–6. Mark 2C. Lot No. 30820 Derby 1970. 32 t.
6357. Mark 2C. Lot No. 30798 Derby 1970. 32 t.
6358–9. Mark 2A. Lot No. 30788 Derby 1968. 31.5 t.
6390. Mark 1. Lot No. 30136 Metro-Cammell 1955. 31.5 t.

6351	(3050, 975435)	**BG**	H	*GN*	EC
6352	(13465, 19465)	**BG**	H	*GN*	BN
6353	(13478, 19478)	**BG**	H	*GN*	EC
6354	(9459)		H	*GN*	BN
6355	(9477)	**BG**	H	*GN*	BN
6356	(9455)	**BG**	H	*GN*	BN
6357	(9443)	**BG**	H	*GN*	BN
6358	(9432)	**BG**	H	*GN*	BN
6359	(9429)	**BG**	H	*GN*	BN
6390	(80723, 92900)		H	*GN*	BN

GF5 (BV) DMU/EMU* BARRIER VEHICLE

Various diagrams. Converted 1992 from BSO or BG*.

6360. Mark 2A. Lot No. 30777 Derby 1967. B4 bogies. 31.5 t.
6361. Mark 2C. Lot No. 30820 Derby 1970. B4 bogies. 32 t.
6364. Mark 1. Lot No. 30039 Derby 1954. BR Mark 1 bogies. 32 t.
6365. Mark 1. Lot No. 30323 Pressed Steel 1957. BR Mark 1 bogies. 32 t.

6360	(9420)		**RR**	P	*P*	NL
6361	(9460)		**RR**	P	*P*	NL
6364	(80565)	*	**RR**	P	*P*	TS
6365	(81296, 84296)	*	**RR**	P	*P*	TS

GS5 (HSBV) HST BARRIER VEHICLE

Dia. GS507. Mark 1. Converted from BG in 1994–5. B4 bogies.

6392. Lot No. 30715 Gloucester 1962. 29.5 t.
6393/6/7. Lot No. 30716 Gloucester 1962. 29.5 t.
6394. Lot No. 30162 Pressed Steel 1956–7. 30.5 t.
6395. Lot No. 30484 Pressed Steel 1958. 30.5 t.
6398/9. Lot No. 30400 Pressed Steel 1957–8. 30.5 t.

6392	(81588, 92183)	**P**	P	*P*	LA
6393	(81609, 92196)	**P**	P	*P*	LA
6394	(80878, 92906)	**P**	P	*P*	NL
6395	(81506, 92148)	**P**	P	*P*	NL
6396	(81606, 92195)	**P**	P	*P*	LA
6397	(81600, 92190)	**P**	P	*P*	NL

6398	(81471, 92126)	**P**	P	*P*	NL
6399	(81367, 92994)	**P**	P	*P*	NL

AG2C (TSOT) OPEN STANDARD (TROLLEY)

Dia. AG201. Mark 2C. Converted from TSO by removal of one seating bay and replacing this by a counter with a space for a trolley. Adjacent toilet removed and converted to steward's washing area/store. Pressure ventilated. –/54 1T. B4 bogies. ETH 4.

Lot No. 30795 Derby 1969–70. 32.5 t.

6523	(5569)	**BG**	WC		CS
6528	(5592)	**M**	WC	*ON*	CS

AG2D (TSOT) OPEN STANDARD (TROLLEY)

Dia. AG202. Mark 2D. Converted from TSO by removal of one seating bay and replacing this by a counter with a space for a trolley. Adjacent toilet removed and converted to steward's washing area/store. Air conditioned. Stones equipment. –/54 1T. B4 bogies. ETH 5.

Lot No. 30822 Derby 1971. 33 t.

6609	(5698)	H	KN
6619	(5655)	H	KN

AN1F (RLO) SLEEPER RECEPTION CAR

Dia. AN101 (AN102*). Mark 2F. Converted from FO, these vehicles consist of pantry, microwave cooking facilities, seating area for passengers, telephone booth and staff toilet. 6703–8 also have a bar. Converted at RTC, Derby (6700), Ilford (6701–5) and Derby (6706–8). Air conditioned. 6700/1/3/5/–8 have Stones equipment and 6702/4 have Temperature Ltd. 26/– 1T. B4 bogies. p. q. d. ETH 5X.

6700–2/4/8. Lot No. 30859 Derby 1973–4. 33.5 t.
6703/5–7. Lot No. 30845 Derby 1973. 33.5 t.

6700	(3347)		**SS**	H	*SR*	IS
6701	(3346)	*		H	*SR*	IS
6702	(3421)	*		H	*SR*	IS
6703	(3308)		**SS**	H	*SR*	IS
6704	(3341)			H	*SR*	IS
6705	(3310, 6430)			H	*SR*	IS
6706	(3283, 6421)			H	*SR*	IS
6707	(3276, 6418)			H	*SR*	IS
6708	(3370)			H	*SR*	IS

AN1D (RMBF) MINIATURE BUFFET CAR

Dia. AN103. Mark 2D. Converted from TSOT by the removal of another seating bay and fitting a proper buffet counter with boiler and microwave oven. Now converted to first class with new seating. Air conditioned. Stones equipment. 30/– 1T. B4 bogies. p. q. d. ETH 5.

Lot No. 30822 Derby 1971. 33 t.

6720	(5622, 6652)	**GW**	H	*GW*	OO
6721	(5627, 6660)		H		ZC
6722	(5736, 6661)	**GW**	H	*GW*	OO
6723	(5641, 6662)		H		ZC
6724	(5721, 6665)	**GW**	H	*GW*	OO

AC2F (TSO) OPEN STANDARD

Dia. AC224. Mark 2F. Renumbered from FO and declassified in 1985–6. Converted 1990 to TSO with mainly unidirectional seating and power-operated sliding doors. Air conditioned. 6800–14 were converted by BREL Derby and have Temperature Ltd. air conditioning. 6815–29 were converted by RFS Industries Doncaster and have Stones air conditioning. –/74 2T. B4 bogies. d. ETH 5X.

r Refurbished vehicles.

6800–07. 6810–12. 6813–14. 6819/22/28. Lot No. 30859 Derby 1973–4. 33 t.
6808–6809. Lot No. 30873 Derby 1974–5. 33.5 t.
6815–18. 6820–21. 6823–27. 6829. Lot No. 30845 Derby 1973. 33 t.

6800	(3323, 6435)	**AR**	H	*AR*	NC
6801	(3349, 6442)		H	*AR*	NC
6802	(3339, 6439)		H	*AR*	NC
6803	(3355, 6443)	**AR**	H	*AR*	NC
6804	(3396, 6449)		H	*AR*	NC
6805	(3324, 6436)	**AR**	H	*AR*	NC
6806	(3342, 6440)	**AR**	H	*AR*	NC
6807	(3423, 6452)		H	*AR*	NC
6808	(3430, 6454)	**AR**	H	*AR*	NC
6809	(3435, 6455)	**AR**	H	*AR*	NC
6810	(3404, 6451)	**AR**	H	*AR*	NC
6811	(3327, 6437)		H	*AR*	NC
6812	(3394, 6448)	**AR**	H	*AR*	NC
6813	(3410, 6463)		H	*AR*	NC
6814	(3422, 6465)	**AR**	H	*AR*	NC
6815	(3282, 6420)	**AR**	H	*AR*	NC
6816	(3316, 6461)	**AR**	H	*AR*	NC
6817	(3311, 6431)		H	*AR*	NC
6818	(3298, 6427)	**AR**	H	*AR*	NC
6819	(3365, 6446)		H	*AR*	NC
6820	(3320, 6434)	**AR**	H	*AR*	NC
6821	(3281, 6458)	**AR**	H	*AR*	NC
6822	(3376, 6447)		H	*AR*	NC
6823	(3289, 6424)		H	*AR*	NC
6824	(3307, 6429)	**AR**	H	*AR*	NC
6825	(3301, 6460)	**AR**	H	*AR*	NC
6826	(3294, 6425)		H	*AR*	NC
6827	(3306, 6428)	**AR**	H	*AR*	NC
6828	(3380, 6464)		H	*AR*	NC
6829	(3288, 6423)	**AR**	H	*AR*	NC

NM51 MERSEYRAIL SANDITE COACH

Dia. NM504. Mark 1. Former Class 501 750 V d.c. third rail EMU driving trailers converted for use as Sandite/de-icing coaches. BR Mark 1 Bogies.

Lot No. 30328 Ashford/Eastleigh 1958. . t.

6910	(75178, 977346)	**MD**	RT	*OD*	BD
6911	(75180, 977348)	**MD**	RT	*OD*	BD

AH2Z (BSOT) OPEN BRAKE STANDARD (MICRO-BUFFET)

Dia. AH203. Mark 2. Converted from BSO by removal of one seating bay and replacing this by a counter with a space for a trolley. Adjacent toilet removed and converted to a steward's washing area/store. –/23 0T. B4 bogies. ETH 4.

Lot No. 30757 Derby 1966. 31 t.

9100	(9405)	v	**RR**	H		LT
9101	(9398)	v	**RR**	BM		TM
9104	(9401)	v	**G**	MH	*ON*	RL
9105	(9404)	v	**RR**	H		LT

AE21 (BSO) OPEN BRAKE STANDARD

Dia. AE201. Mark 1. –/39 1T. BR Mark 1 bogies. ETH 3.

Lot No. 30170 Doncaster 1955–6. 34 t.

9227	xk	**M**	SP	*ON*	BT
9274	v	**CH**	NY	*ON*	NY

AE2Z (BSO) OPEN BRAKE STANDARD

Dia. AE203. Mark 2. These vehicles use the same body shell as the Mark 2 BFK and have first class seat spacing and wider tables. Pressure ventilated. –/31 1T. B4 bogies. ETH 4.

Lot No. 30757 Derby 1966. 31.5 t.

9385	v	**LN**	H	LT
9388	v	**LN**	H	LT

AE2A (BSO) OPEN BRAKE STANDARD

Dia. AE204. Mark 2A. These vehicles use the same body shell as the Mark 2 BFK and have first class seat spacing and wider tables. Pressure ventilated. –/31 1T. B4 bogies. ETH 4.

9417–24. Lot No. 30777 Derby 1970. 31.5 t.
9428–35. Lot No. 30820 Derby 1970. 31.5 t.

9417	**FT**	H	*CA*	CF
9418	**RR**	H		PY
9421	**RR**	H		CP
9424	**RR**	H		CP
9428	**DR**	DR	*DR*	PY
9431	**RR**	H		PY
9434	**RR**	H		ZN
9435	**RR**	H		KN

AE2C (BSO) OPEN BRAKE STANDARD

Dia. AE205. Mark 2C. Pressure ventilated. –/31 1T. B4 bogies. ETH 4.

Lot No. 30798 Derby 1970. 32 t.

Non-Standard Livery: 9440 is in Royal blue with white lining.

9440	d	**0**	WC	*WW*	CF
9448	d	**M**	WC	*WW*	CF

AE2D (BSO) OPEN BRAKE STANDARD

Dia. AE206. Mark 2D. Air conditioned (Stones). –/31 1T. B4 bogies. pg. ETH 5.

r Refurbished with new green seat trim.
s Refurbished with new seats.

Lot No. 30824 Derby 1971. 33 t.

9479	dr		H	*VX*	MA
9480	d		H	*GW*	OO
9481	d		H	*GW*	LA
9483			H		PY
9484	d		H		LT
9485			H		LT
9486			H		PY
9488	ds	**GW**	H	*GW*	OO
9489	dr	**V**	H	*VX*	MA
9490	d		H		ZC
9492	d		H	*GW*	LA
9493	d		H		ZC
9494	ds	**GW**	H	*GW*	OO

AE2E (BSO) OPEN BRAKE STANDARD

Dia. AE207. Mark 2E. Air conditioned (Stones). –/32 1T. B4 bogies. d. pg. ETH 5.

Lot No. 30838 Derby 1972. 33 t.

r Refurbished with new green seat trim.
s Refurbished with modified design of seat headrest and new green seat trim.

9496	r		H	*VX*	MA
9497	r		H	*VX*	MA
9498	r	**V**	H	*VX*	MA
9500	r		H	*VX*	MA
9501			H	*GW*	OO
9502	s	**V**	H	*VX*	MA
9503	s	**V**	H	*VX*	MA
9504	s	**V**	H	*VX*	MA
9505	s		H	*VX*	MA
9506	s	**V**	H	*VX*	MA
9507	s	**V**	H	*VX*	MA
9508	s	**V**	H	*VX*	MA
9509	s	**V**	H	*VX*	MA

AE2F (BSO) OPEN BRAKE STANDARD

Dia. AE208. Mark 2F. Air conditioned (Temperature Ltd.). All now refurbished with power-operated vestibule doors, new panels and seat trim. All now further refurbished with green seat trim. –/32 1T. B4 bogies. d. pg. ETH5X.

Lot No. 30861 Derby 1974. 34 t.

9513	**V**	H	*VX*	MA
9516	**V**	H	*VX*	MA
9520	**V**	H	*VX*	MA
9521	**V**	H	*VX*	MA

9522		**V**	H	*VX*	MA
9523		**V**	H	*VX*	MA
9524	n	**V**	H	*VX*	MA
9525		**V**	H	*VX*	MA
9526			H	*VX*	MA
9527		**V**	H	*VX*	MA
9529		**V**	H	*VX*	MA
9531		**V**	H	*VX*	MA
9537	n	**V**	H	*VX*	MA
9538		**V**	H	*VX*	MA
9539		**V**	H	*VX*	MA

AF2F (DBSO) DRIVING OPEN BRAKE STANDARD

Dia. AF201. Mark 2F. Air conditioned (Temperature Ltd.). Push & pull (t.d.m. system). Converted from BSO, these vehicles originally had half cabs at the brake end. They have since been refurbished and have had their cabs widened and the cab-end gangways removed. –/31 1W 1T. B4 bogies. d. pg. Cowcatchers. ETH 5X.

9701–9710. Lot No. 30861 Derby 1974. Converted Glasgow 1979. Disc brakes. 34 t.
9711–9713. Lot No. 30861 Derby 1974. Converted Glasgow 1985. 34 t.
9714. Lot No. 30861 Derby 1974. Converted Glasgow 1986. Disc brakes. 34 t.

9701	(9528)		H	*AR*	NC
9702	(9510)	**AR**	H	*AR*	NC
9703	(9517)	**AR**	H	*AR*	NC
9704	(9512)	**AR**	H	*AR*	NC
9705	(9519)		H	*AR*	NC
9707	(9511)	**AR**	H	*AR*	NC
9708	(9530)	**AR**	H	*AR*	NC
9709	(9515)	**AR**	H	*AR*	NC
9710	(9518)	**AR**	H	*AR*	NC
9711	(9532)	**AR**	H	*AR*	NC
9712	(9534)	**AR**	H	*AR*	NC
9713	(9535)	**AR**	H	*AR*	NC
9714	(9536)	**AR**	H	*AR*	NC

AE4E (BUO) UNCLASSIFIED OPEN BRAKE

Dia. AE401. Mark 2E. Converted from TSO for use on Anglo-Scottish overnight services by Railcare, Wolverton. Air conditioned. Stones equipment. Require at least 800 V train supply. B4 bogies. d. –/31 1T. B4 bogies. ETH 4X.

9801–9803. Lot No. 30837 Derby 1972. 33.5 t.
9804–9810. Lot No. 30844 Derby 1972–3. 33.5 t.

9800	(5751)	H	ZN
9801	(5760)	H	ZN
9802	(5772)	H	ZN
9803	(5799)	H	ZN
9804	(5826)	H	ZN
9805	(5833)	H	ZN
9806	(5840)	H	ZN
9807	(5851)	H	ZN
9808	(5871)	H	ZN
9809	(5890)	H	ZN
9810	(5892)	H	ZN

AJ1G (RFM) RESTAURANT BUFFET FIRST (MODULAR)

Dia. AJ103 (10200/1 are Dia. AJ101). Mark 3A. Air conditioned. Converted from HST TRFKs, RFBs and FOs. 18/– plus two seats for staff use. (24/– *). BT10 bogies. p. q. d. ETH 14X.

10200–10211. Lot No. 30884 Derby 1977.
10212–10229. Lot No. 30878 Derby 1975–6. 39.8 t.

10230–10260. Lot No. 30890 Derby 1979. 39.8 t.

r Refurbished with table lamps and new burgundy seat trim.

10200	(40519)	*		P	*AR*	NC
10201	(40520)	r	**V**	P	*VW*	OY
10202	(40504)	r	**V**	P	*VW*	MA
10203	(40506)	*	**AR**	P	*AR*	NC
10204	(40502)	r	**V**	P	*VW*	MA
10205	(40503)	r	**V**	P	*VW*	OY
10206	(40507)	r	**V**	P	*VW*	MA
10207	(40516)	r	**V**	P	*VW*	PC
10208	(40517)	r	**V**	P	*VW*	MA
10209	(40508)	r	**V**	P	*VW*	PC
10210	(40509)	r	**V**	P	*VW*	PC
10211	(40510)	r	**V**	P	*VW*	PC
10212	(11049)	r	**V**	P	*VW*	MA
10213	(11050)	r	**V**	P	*VW*	MA
10214	(11034)	*	**AR**	P	*AR*	NC
10215	(11032)	r	**V**	P	*VW*	PC
10216	(11041)	*	**AR**	P	*AR*	NC
10217	(11051)	r	**V**	P	*VW*	MA
10218	(11053)	r	**V**	P	*VW*	MA
10219	(11047)	r	**V**	P	*VW*	PC
10220	(11056)	r	**V**	P	*VW*	OY
10221	(11012)	r	**V**	P	*VW*	PC
10222	(11063)	r	**V**	P	*VW*	MA
10223	(11043)	*	**AR**	P	*AR*	NC
10224	(11062)	r	**V**	P	*VW*	MA
10225	(11014)	r		P	*VW*	OY
10226	(11015)	r	**V**	P	*VW*	MA
10227	(11057)	r	**V**	P	*VW*	PC
10228	(11035)	*	**AR**	P	*AR*	NC
10229	(11059)	r	**V**	P	*VW*	MA
10230	(10021)	r	**V**	P	*VW*	PC
10231	(10016)	r	**V**	P	*VW*	OY
10232	(10027)	r		P	*VW*	OY
10233	(10013)	r	**V**	P	*VW*	PC
10234	(10004)	r	**V**	P	*VW*	PC
10235	(10015)	r		P	*VW*	OY
10236	(10018)			P	*VW*	PC
10237	(10022)	r	**V**	P	*VW*	MA
10238	(10017)	r	**V**	P	*VW*	OY
10240	(10003)	r	**V**	P	*VW*	OY
10241	(10009)	*	**AR**	P	*AR*	NC
10242	(10002)	r		P	*VW*	OY
10245	(10019)	r	**V**	P	*VW*	PC
10246	(10014)	r	**V**	P	*VW*	PC
10247	(10011)	*	**AR**	P	*AR*	NC
10248	(10005)	r	**V**	P	*VW*	OY
10249	(10012)	r	**V**	P	*VW*	PC
10250	(10020)	r	**V**	P	*VW*	OY
10251	(10024)	r	**V**	P	*VW*	OY
10252	(10008)	r	**V**	P	*VW*	OY
10253	(10026)	r	**V**	P	*VW*	PC
10254	(10006)	r	**V**	P	*VW*	PC
10255	(10010)	r	**V**	P	*VW*	OY
10256	(10028)	r	**V**	P	*VW*	PC
10257	(10007)	r	**V**	P	*VW*	PC
10258	(10023)			P	*VW*	MA
10259	(10025)	r	**V**	P	*VW*	OY
10260	(10001)	r	**V**	P	*VW*	MA

AJ1J (RFM) RESTAURANT BUFFET FIRST (MODULAR)

Dia. AJ105. Mark 4. Air conditioned. 20/– 1T. BT41 bogies. ETH 6X.

Lot No. 31045 Metro-Cammell 1989 onwards. 45.5 t.

10300	**GN**	H	*GN*	BN
10301	**GN**	H	*GN*	BN
10302	**GN**	H	*GN*	BN
10303	**GN**	H	*GN*	BN
10304	**GN**	H	*GN*	BN
10305	**GN**	H	*GN*	BN
10306	**GN**	H	*GN*	BN
10307	**GN**	H	*GN*	BN
10308	**GN**	H	*GN*	BN
10309	**GN**	H	*GN*	BN
10310	**GN**	H	*GN*	BN
10311	**GN**	H	*GN*	BN
10312	**GN**	H	*GN*	BN
10313	**GN**	H	*GN*	BN
10314	**GN**	H	*GN*	BN
10315	**GN**	H	*GN*	BN
10316	**GN**	H	*GN*	BN
10317	**GN**	H	*GN*	BN
10318	**GN**	H	*GN*	BN
10319	**GN**	H	*GN*	BN
10320	**GN**	H	*GN*	BN
10321	**GN**	H	*GN*	BN
10322	**GN**	H	*GN*	BN
10323	**GN**	H	*GN*	BN

10324	**GN**	H *GN*	BN	10329	**GN**	H *GN*	BN
10325	**GN**	H *GN*	BN	10330	**GN**	H *GN*	BN
10326	**GN**	H *GN*	BN	10331	**GN**	H *GN*	BN
10327	**GN**	H *GN*	BN	10332	**GN**	H *GN*	BN
10328	**GN**	H *GN*	BN	10333	**GN**	H *GN*	BN

AU4G (SLEP) SLEEPING CAR WITH PANTRY

Dia. AU401. Mark 3A. Air conditioned. Retention toilets. 12 compartments with a fixed lower berth and a hinged upper berth, plus an attendants compartment. 2T BT10 bogies. ETH 7X.

Lot No. 30960 Derby 1981–3. 41 t.

10500			SS		ZC				
10501	d		P	*SR*	IS				
10502	d		P	*SR*	IS				
10503			SS		ZC				
10504	d		P	*SR*	IS				
10506	d		P	*SR*	IS				
10507	d		P	*SR*	IS				
10508	d		P	*SR*	IS				
10510	d		P	*SR*	IS				
10512	d		P		ZG				
10513	d		P	*SR*	IS				
10514			SS		ZC				
10515	d		P	*SR*	IS				
10516	d		P	*SR*	IS				
10519	d		P	*SR*	IS				
10520	d		P	*SR*	IS				
10522	d		P	*SR*	IS				
10523	d		P	*SR*	IS				
10526	d		P	*SR*	IS				
10527	d	**SS**	P	*SR*	IS				
10529	d		P	*SR*	IS				
10530	d		P		ZD				
10531	d		P	*SR*	IS				
10532	d	**GW**	P	*GW*	PZ				
10533			P		ZD				
10534	d	**GW**	P	*GW*	PZ				
10535	d		P		ZD				
10536	d		P		KN				
10537	d		P		ZD				
10538	d		P		KN				
10539	d		P		KN				
10540	d		P		ZD				
10542	d		P	*SR*	IS				
10543	d		P	*SR*	IS				
10544	d		P	*SR*	IS				
10546			P		ZD				
10547	d		P	*SR*	IS				
10548	d		P	*SR*	IS				

10549	d		P		ZD
10550	d		P		ZD
10551	d	**SS**	P	*SR*	IS
10553	d		P	*SR*	IS
10554	d		P		ZD
10555	d		P		KN
10557	d		P		ZD
10558	d		P		ZH
10559	d		P		KN
10560	d		P		ZD
10561	d		P	*SR*	IS
10562	d		P	*SR*	IS
10563	d	**GW**	P	*GW*	PZ
10565	d		P	*SR*	IS
10566	d		P		ZD
10567			P		ZG
10569	d	**PC**	VS	*ON*	SL
10570			P		KN
10571			SS		BN
10572	d		P		ZD
10574			CN		FK
10575			SS		ZC
10577		**BG**	P		ZD
10578			P		KN
10579		**BG**	P		KN
10580	d*	**SS**	P	*SR*	IS
10582	d		P		ZD
10583	d	**GW**	P	*GW*	PZ
10584	d	**GW**	P	*GW*	PZ
10586	d		P		KN
10588	d*	**GW**	P	*GW*	PZ
10589	d	**GW**	P	*GW*	PZ
10590	d	**GW**	P	*GW*	PZ
10592			P		KN
10593	d		P		KN
10594	d	**GW**	P	*GW*	PZ
10595		**BG**	P		KN
10596	d		P		KN

10597 d		P	*SR*	IS
10598 d		P	*SR*	IS
10599		P		KN
10600 d		P	*SR*	IS
10601		P		ZD
10602		P		ZD
10604		P		ZD
10605 d		P	*SR*	IS
10606		P		KN
10607 d		P	*SR*	IS
10608	**BG**	P		ZN
10609	**BG**	P		ZG
10610 d		P	*SR*	IS
10612 d	**GW**	P	*GW*	PZ
10613 d		P	*SR*	IS
10614 d	**SS**	P	*SR*	IS
10616 d	**GW**	P	*GW*	PZ
10617 d	**SS**	P	*SR*	IS

AS4G (SLE/SLED*) SLEEPING CAR

Dia. AS403 (AS404*). Mark 3A. Air conditioned. Retention toilets. 13 compartments with a fixed lower berth and a hinged upper berth (* 11 compartments with a fixed lower berth and a hinged upper berth + one compartment for a disabled person). 2T. BT10 bogies. ETH 6X.

Notes:

10664/7/9/76/7/81/94/5/8/721 were sold to the Danish State Railways (DSB) when they were numbered in the UIC system. They have since been purchased by Angel Train Contracts and returned to Britain.

10729 is leased to Venice Simplon-Orient Express.

Lot No. 30961 Derby 1980–4. 43.5 t.

10646 d		CN		FK
10647 d		P		KN
10648 d*		P	*SR*	IS
10649 d		P		KN
10650 d*		P	*SR*	IS
10651 d		P		ZD
10653 d		P		ZD
10654 d		P		ZD
10655		SS		ZC
10656		P		KN
10657		SS		ZC
10658 d		P		KN
10660 d		P		ZD
10662		P		ZD
10663 d		P	*SR*	IS
10664	**DS**	A		ZC
10665	**BG**	P		ZG
10666 d		P	*SR*	IS
10667	**DS**	A		ZC
10668 d		P		ZD
10669	**DS**	A		ZC
10670		P		KN
10672 d		P		ZG
10674 d		P		ZG
10675 d		P	*SR*	IS
10676	**DS**	A		ZC
10677	**DS**	A		ZC
10678	**BG**	P		KN
10679	**BG**	P		KN
10680 d*		P	*SR*	IS
10681	**DS**	A		ZC
10682 d		P		ZD
10683 d	**SS**	P	*SR*	IS
10684	**BG**	P		KN
10685 d		P		IS
10686 d		P		ZD
10687 d		P		ZD
10688 d*	**SS**	P	*SR*	IS
10689 d		P	*SR*	IS
10690 d		P	*SR*	IS
10691 d		P		ZD
10692 d		P		ZD
10693 d		P	*SR*	IS
10694	**DS**	A		ZC
10695	**DS**	A		ZC
10696 d		P		KN
10697 d		P		KN
10698	**DS**	A		ZC
10699 d*		P	*SR*	IS
10700	**BG**	P		KN
10701 d		P		KN
10702		SS		ZC
10703 d		P	*SR*	IS
10704 d		P	*AE*	ZA

No.	Livery	Owner	Operator	Depot
10706 d		P	SR	IS
10707		P		ZG
10708 d		P		ZD
10709 d		P		ZD
10710 d		P		KN
10711 d		P		ZD
10712 d		P		ZD
10713 d		P		ZD
10714 d*		P	SR	IS
10715 d		P		ZD
10716 d		P		ZD
10717 d		P		ZD
10718 d*		P	SR	IS
10719 d*		P	SR	IS
10720		P		KN
10721	**DS**	A		ZC
10722 d*		P	SR	IS
10723 d*		P	SR	IS
10724		SS		FK
10725		SS		ZC
10726		SS		ZC
10727		SS		ZC
10728		P		ZN
10729	**RB**	SS	ON	CP
10730 d		P		ZD
10731 d		P		KN
10732 d		P		KN

AD1G (FO) OPEN FIRST

Dia. AD108. Mark 3A. Air conditioned. 48/– 2T (* 48/– 1T 1TD). BT10 bogies. d. ETH 6X.

11005–7 were open composites 11905–7 for a time.

Lot No. 30878 Derby 1975–6. 34.3 t.

r Refurbished with table lamps and burgundy seat trim.

No.	Livery	Owner	Operator	Depot
11005 r	**V**	P	*VW*	PC
11006 r	**V**	P	*VW*	PC
11007 r	**V**	P	*VW*	PC
11011 r*	**V**	P	*VW*	MA
11013 r	**V**	P	*VW*	PC
11016 r	**V**	P	*VW*	PC
11017		P	*VW*	PC
11018 r	**V**	P	*VW*	MA
11019 r	**V**	P	*VW*	PC
11020 r	**V**	P	*VW*	MA
11021 r	**V**	P	*VW*	PC
11023 r	**V**	P	*VW*	PC
11024 r	**V**	P	*VW*	MA
11026 r	**V**	P	*VW*	MA
11027 r	**V**	P	*VW*	MA
11028 r	**V**	P	*VW*	MA
11029 r	**V**	P	*VW*	MA
11030 r	**V**	P	*VW*	MA
11031 r	**V**	P	*VW*	MA
11033 r	**V**	P	*VW*	PC
11036 r	**V**	P	*VW*	MA
11037 r	**V**	P	*VW*	PC
11038 r	**V**	P	*VW*	PC
11039 r	**V**	P	*VW*	PC
11040 r	**V**	P	*VW*	MA
11042 r	**V**	P	*VW*	MA
11044 r	**V**	P	*VW*	MA
11045 r	**V**	P	*VW*	PC
11046 r	**V**	P	*VW*	PC
11048 r	**V**	P	*VW*	MA
11052 r	**V**	P	*VW*	MA
11054 r	**V**	P	*VW*	PC
11055		P	*VW*	PC
11058 r	**V**	P	*VW*	MA
11060 r	**V**	P	*VW*	PC

AD1H (FO) OPEN FIRST

Dia. AD109. Mark 3B. Air conditioned. Inter-City 80 seats. 48/– 2T. BT10 bogies. d. ETH 6X.

Lot No. 30982 Derby 1985. 36.5 t.

r Refurbished with table lamps and burgundy seat trim..

11064	r	**V**	P	*VW*	MA
11065	r	**V**	P	*VW*	PC
11066	r	**V**	P	*VW*	MA
11067	r	**V**	P	*VW*	PC
11068			P	*VW*	MA
11069	r	**V**	P	*VW*	PC
11070	r	**V**	P	*VW*	MA
11071	r	**V**	P	*VW*	PC
11072	r	**V**	P	*VW*	PC
11073	r	**V**	P	*VW*	MA
11074	r	**V**	P	*VW*	MA
11075	r	**V**	P	*VW*	MA
11076	r	**V**	P	*VW*	PC
11077	r	**V**	P	*VW*	MA
11078	r	**V**	P	*VW*	PC
11079	r	**V**	P	*VW*	MA
11080	r	**V**	P	*VW*	MA
11081	r	**V**	P	*VW*	PC
11082	r	**V**	P	*VW*	PC
11083	pr	**V**	P	*VW*	MA
11084	pr	**V**	P	*VW*	MA
11085	pr	**V**	P	*VW*	MA
11086	pr	**V**	P	*VW*	PC
11087	pr	**V**	P	*VW*	MA
11088	pr	**V**	P	*VW*	PC
11089	pr	**V**	P	*VW*	PC
11090	p		P	*VW*	MA
11091	pr	**V**	P	*VW*	MA
11092	p		P	*VW*	MA
11093	pr	**V**	P	*VW*	PC
11094	pr	**V**	P	*VW*	MA
11095	pr	**V**	P	*VW*	MA
11096	pr	**V**	P	*VW*	PC
11097	pr	**V**	P	*VW*	MA
11098	pr	**V**	P	*VW*	PC
11099	pr	**V**	P	*VW*	PC
11100	pr	**V**	P	*VW*	PC
11101	pr	**V**	P	*VW*	MA

AD1J (FO) OPEN FIRST

Dia. AD111. Mark 4. Air conditioned. 46/– 1T. BT41 bogies. ETH 6X.

11264–71 were cancelled.

Lot No. 31046 Metro-Cammell 1989–92. 39.7 t.

11200		**GN**	H	*GN*	BN
11201	p	**GN**	H	*GN*	BN
11202		**GN**	H	*GN*	BN
11203	p	**GN**	H	*GN*	BN
11204	p	**GN**	H	*GN*	BN
11205		**GN**	H	*GN*	BN
11206		**GN**	H	*GN*	BN
11207	p	**GN**	H	*GN*	BN
11208		**GN**	H	*GN*	BN
11209		**GN**	H	*GN*	BN
11210		**GN**	H	*GN*	BN
11211	p	**GN**	H	*GN*	BN
11212		**GN**	H	*GN*	BN
11213	p	**GN**	H	*GN*	BN
11214	p	**GN**	H	*GN*	BN
11215		**GN**	H	*GN*	BN
11216		**GN**	H	*GN*	BN
11217	p	**GN**	H	*GN*	BN
11218		**GN**	H	*GN*	BN
11219	p	**GN**	H	*GN*	BN
11220		**GN**	H	*GN*	BN
11221	p	**GN**	H	*GN*	BN
11222	p	**GN**	H	*GN*	BN
11223		**GN**	H	*GN*	BN
11224		**GN**	H	*GN*	BN
11225	p	**GN**	H	*GN*	BN
11226		**GN**	H	*GN*	BN
11227	p	**GN**	H	*GN*	BN
11228	p	**GN**	H	*GN*	BN
11229	p	**GN**	H	*GN*	BN
11230		**GN**	H	*GN*	BN
11231	p	**GN**	H	*GN*	BN
11232		**GN**	H	*GN*	BN
11233	p	**GN**	H	*GN*	BN
11234		**GN**	H	*GN*	BN
11235	p	**GN**	H	*GN*	BN
11236		**GN**	H	*GN*	BN
11237	p	**GN**	H	*GN*	BN
11238		**GN**	H	*GN*	BN
11239	p	**GN**	H	*GN*	BN
11240		**GN**	H	*GN*	BN
11241		**GN**	H	*GN*	BN
11242	p	**GN**	H	*GN*	BN
11243	p	**GN**	H	*GN*	BN
11244		**GN**	H	*GN*	BN
11245	p	**GN**	H	*GN*	BN
11246	p	**GN**	H	*GN*	BN
11247	p	**GN**	H	*GN*	BN

11248	**GN**	H	*GN*	BN
11249 p	**GN**	H	*GN*	BN
11250	**GN**	H	*GN*	BN
11251 p	**GN**	H	*GN*	BN
11252	**GN**	H	*GN*	BN
11253 p	**GN**	H	*GN*	BN
11254	**GN**	H	*GN*	BN
11255 p	**GN**	H	*GN*	BN
11256	**GN**	H	*GN*	BN
11257 p	**GN**	H	*GN*	BN
11258	**GN**	H	*GN*	BN
11259 p	**GN**	H	*GN*	BN
11260	**GN**	H	*GN*	BN
11261 p	**GN**	H	*GN*	BN
11262	**GN**	H	*GN*	BN
11263 p	**GN**	H	*GN*	BN
11272	**GN**	H	*GN*	BN
11273	**GN**	H	*GN*	BN
11274	**GN**	H	*GN*	BN
11275	**GN**	H	*GN*	BN
11276	**GN**	H	*GN*	BN

AC2G (TSO) OPEN STANDARD

Dia. AC213 (AC220 z). Mark 3A. Air conditioned. All now refurbished with modified seat backs and new layout. –/76 2T (w –/74 2T, z –/74 1T 1TD). BT10 (§ BREL T4) bogies. d. ETH 6X.

Note: 12169–72 were converted from open composites 11908–10/22, formerly FOs 11008–10/22.

Lot No. 30877 Derby 1975–7. 34.3 t.

r Further refurbished with light blue seat trim.
s Further refurbished with light blue seat trim and two wheelchair spaces. –/70 2T 2W (z –/70 1TD 1T 2W).

12004 r	**V**	P	*VW*	MA
12005 r	**V**	P	*VW*	PC
12007 r	**V**	P	*VW*	MA
12008 r	**V**	P	*VW*	MA
12009 r	**V**	P	*VW*	MA
12010 r	**V**	P	*VW*	MA
12011 r	**V**	P	*VW*	PC
12012 r	**V**	P	*VW*	PC
12013 r	**V**	P	*VW*	MA
12014 r	**V**	P	*VW*	PC
12015 r	**V**	P	*VW*	PC
12016 r	**V**	P	*VW*	PC
12017 r	**V**	P	*VW*	MA
12019 r	**V**	P	*VW*	PC
12020 r	**V**	P	*VW*	MA
12021 r	**V**	P	*VW*	PC
12022 r	**V**	P	*VW*	MA
12023 r	**V**	P	*VW*	PC
12024 s	**V**	P	*VW*	PC
12025 r	**V**	P	*VW*	MA
12026 r	**V**	P	*VW*	PC
12027 r	**V**	P	*VW*	MA
12028 r	**V**	P	*VW*	MA
12029		P	*VW*	MA
12030 r	**V**	P	*VW*	PC
12031 r	**V**	P	*VW*	PC
12032 r	**V**	P	*VW*	PC
12033 sz	**V**	P	*VW*	MA
12034 r	**V**	P	*VW*	MA
12035 r	**V**	P	*VW*	PC
12036 s	**V**	P	*VW*	PC
12037 r	**V**	P	*VW*	PC
12038 r	**V**	P	*VW*	PC
12040 r	**V**	P	*VW*	PC
12041 r	**V**	P	*VW*	PC
12042 s	**V**	P	*VW*	PC
12043 r	**V**	P	*VW*	MA
12044 r	**V**	P	*VW*	MA
12045 r	**V**	P	*VW*	MA
12046 r	**V**	P	*VW*	PC
12047 z		P	*VW*	PC
12048 r	**V**	P	*VW*	PC
12049		P	*VW*	PC
12050 s	**V**	P	*VW*	PC
12051 r	**V**	P	*VW*	PC
12052 r	**V**	P	*VW*	MA
12053 r	**V**	P	*VW*	MA
12054 s	**V**	P	*VW*	MA
12055 r	**V**	P	*VW*	PC
12056 r	**V**	P	*VW*	PC
12057 r	**V**	P	*VW*	MA
12058 r	**V**	P	*VW*	PC

12059	w		P	*VW*	MA
12060	r	**V**	P	*VW*	PC
12061	s	**V**	P	*VW*	PC
12062	r	**V**	P	*VW*	PC
12063	r	**V**	P	*VW*	MA
12064			P	*VW*	MA
12065	r	**V**	P	*VW*	MA
12066	r	**V**	P	*VW*	MA
12067	r	**V**	P	*VW*	MA
12068			P	*VW*	MA
12069	r	**V**	P	*VW*	MA
12070	r	**V**	P	*VW*	PC
12071	r	**V**	P	*VW*	PC
12072	r	**V**	P	*VW*	MA
12073	r	**V**	P	*VW*	MA
12075	r	**V**	P	*VW*	PC
12076	r	**V**	P	*VW*	PC
12077	r	**V**	P	*VW*	PC
12078	r	**V**	P	*VW*	MA
12079	r	**V**	P	*VW*	PC
12080	r	**V**	P	*VW*	PC
12081	r	**V**	P	*VW*	PC
12082	r	**V**	P	*VW*	PC
12083	r	**V**	P	*VW*	MA
12084			P	*VW*	MA
12085	s	**V**	P	*VW*	MA
12086	s	**V**	P	*VW*	MA
12087	s	**V**	P	*VW*	PC
12088	sz	**V**	P	*VW*	PC
12089	r	**V**	P	*VW*	MA
12090			P	*VW*	MA
12091	r	**V**	P	*VW*	PC
12092	r	**V**	P	*VW*	MA
12093	r	**V**	P	*VW*	PC
12094	r	**V**	P	*VW*	PC
12095	r	**V**	P	*VW*	MA
12096	r	**V**	P	*VW*	PC
12097			P	*VW*	PC
12098	r	**V**	P	*VW*	MA
12099	r	**V**	P	*VW*	PC
12100	sz	**V**	P	*VW*	PC
12101	s	**V**	P	*VW*	PC
12102	r	**V**	P	*VW*	PC
12103	s	**V**	P	*VW*	MA
12104	r	**V**	P	*VW*	MA
12105	r	**V**	P	*VW*	PC
12106	r	**V**	P	*VW*	MA
12107	r	**V**	P	*VW*	PC
12108	s	**V**	P	*VW*	MA
12109	s	**V**	P	*VW*	MA
12110	r	**V**	P	*VW*	MA
12111	r	**V**	P	*VW*	MA
12112	sz	**V**	P	*VW*	MA
12113	r	**V**	P	*VW*	MA
12114	r	**V**	P	*VW*	PC
12115	r	**V**	P	*VW*	PC
12116	r	**V**	P	*VW*	PC
12117	r	**V**	P	*VW*	MA
12118	r	**V**	P	*VW*	MA
12119	r	**V**	P	*VW*	MA
12120	r	**V**	P	*VW*	MA
12121	r	**V**	P	*VW*	PC
12122	sz	**V**	P	*VW*	MA
12123	r	**V**	P	*VW*	PC
12124	r	**V**′	P	*VW*	MA
12125	r	**V**	P	*VW*	MA
12126	r	**V**	P	*VW*	MA
12127	r	**V**	P	*VW*	PC
12128	s	**V**	P	*VW*	MA
12129	r	**V**	P	*VW*	MA
12130	r	**V**	P	*VW*	MA
12131	r	**V**	P	*VW*	PC
12132	r	**V**	P	*VW*	MA
12133	r	**V**	P	*VW*	MA
12134	r	**V**	P	*VW*	PC
12135			P	*VW*	PC
12136	r	**V**	P	*VW*	PC
12137	r	**V**	P	*VW*	PC
12138	r	**V**	P	*VW*	PC
12139	r	**V**	P	*VW*	MA
12140	sz*	**V**	P	*VW*	PC
12141	r	**V**	P	*VW*	PC
12142	s	**V**	P	*VW*	PC
12143	r	**V**	P	*VW*	PC
12144	s	**V**	P	*VW*	PC
12145	r	**V**	P	*VW*	MA
12146	r	**V**	P	*VW*	PC
12147	r	**V**	P	*VW*	PC
12148	r	**V**	P	*VW*	PC
12149	r	**V**	P	*VW*	PC
12150	r	**V**	P	*VW*	PC
12151	r	**V**	P	*VW*	PC
12152	r	**V**	P	*VW*	PC
12153	r	**V**	P	*VW*	PC
12154	r	**V**	P	*VW*	MA
12155	s	**V**	P	*VW*	PC
12156	r	**V**	P	*VW*	MA
12157	r	**V**	P	*VW*	MA
12158	r	**V**	P	*VW*	PC
12159	r	**V**	P	*VW*	PC
12160	s	**V**	P	*VW*	PC
12161	sz	**V**	P	*VW*	MA

12163 r	**V**	P	*VW*	MA	12168 s	**V**	P	*VW*	PC
12164 r	**V**	P	*VW*	PC	12169 s	**V**	P	*VW*	MA
12165 r	**V**	P	*VW*	MA	12170 s	**V**	P	*VW*	MA
12166 r	**V**	P	*VW*	PC	12171 s	**V**	P	*VW*	PC
12167 r	**V**	P	*VW*	PC	12172 s	**V**	P	*VW*	PC

AI2J (TSOE) OPEN STANDARD (END)

Dia. AI201. Mark 4. Air conditioned. –/74 2T. BT41 bogies. ETH 6X.

Lot No. 31047 Metro-Cammell 1989–91. 39.5 t.

12232 was converted from the original 12405.

12200	**GN**	H	*GN*	BN	12216	**GN**	H	*GN*	BN
12201	**GN**	H	*GN*	BN	12217	**GN**	H	*GN*	BN
12202	**GN**	H	*GN*	BN	12218	**GN**	H	*GN*	BN
12203	**GN**	H	*GN*	BN	12219	**GN**	H	*GN*	BN
12204	**GN**	H	*GN*	BN	12220	**GN**	H	*GN*	BN
12205	**GN**	H	*GN*	BN	12222	**GN**	H	*GN*	BN
12206	**GN**	H	*GN*	BN	12223	**GN**	H	*GN*	BN
12207	**GN**	H	*GN*	BN	12224	**GN**	H	*GN*	BN
12208	**GN**	H	*GN*	BN	12225	**GN**	H	*GN*	BN
12209	**GN**	H	*GN*	BN	12226	**GN**	H	*GN*	BN
12210	**GN**	H	*GN*	BN	12227	**GN**	H	*GN*	BN
12211	**GN**	H	*GN*	BN	12228	**GN**	H	*GN*	BN
12212	**GN**	H	*GN*	BN	12229	**GN**	H	*GN*	BN
12213	**GN**	H	*GN*	BN	12230	**GN**	H	*GN*	BN
12214	**GN**	H	*GN*	BN	12231	**GN**	H	*GN*	BN
12215	**GN**	H	*GN*	BN	12232	**GN**	H	*GN*	BN

AL2J (TSOD) OPEN STANDARD (DISABLED ACCESS)

Dia. AL201. Mark 4. Air conditioned. –/72 1TD 1W. BT41 bogies. p. ETH 6X.

Lot No. 31048 Metro-Cammell 1989–91. 39.4 t.

12300	**GN**	H	*GN*	BN	12316	**GN**	H	*GN*	BN
12301	**GN**	H	*GN*	BN	12317	**GN**	H	*GN*	BN
12302	**GN**	H	*GN*	BN	12318	**GN**	H	*GN*	BN
12303	**GN**	H	*GN*	BN	12319	**GN**	H	*GN*	BN
12304	**GN**	H	*GN*	BN	12320	**GN**	H	*GN*	BN
12305	**GN**	H	*GN*	BN	12321	**GN**	H	*GN*	BN
12306	**GN**	H	*GN*	BN	12322	**GN**	H	*GN*	BN
12307	**GN**	H	*GN*	BN	12323	**GN**	H	*GN*	BN
12308	**GN**	H	*GN*	BN	12324	**GN**	H	*GN*	BN
12309	**GN**	H	*GN*	BN	12325	**GN**	H	*GN*	BN
12310	**GN**	H	*GN*	BN	12326	**GN**	H	*GN*	BN
12311	**GN**	H	*GN*	BN	12327	**GN**	H	*GN*	BN
12312	**GN**	H	*GN*	BN	12328	**GN**	H	*GN*	BN
12313	**GN**	H	*GN*	BN	12329	**GN**	H	*GN*	BN
12314	**GN**	H	*GN*	BN	12330	**GN**	H	*GN*	BN
12315	**GN**	H	*GN*	BN					

AC2J (TSO) OPEN STANDARD

Dia. AC214. Mark 4. Air conditioned. –/74 2T. BT41 bogies. ETH 6X.

Lot No. 31049 Metro-Cammell 1989 onwards. 39.9 t.

12405 is the second coach to carry that number. It was built from the bodyshell originally intended for 12221. The original 12405 is now 12232. 12490–12512 were cancelled.

12400	**GN**	H *GN*	BN
12401	**GN**	H *GN*	BN
12402	**GN**	H *GN*	BN
12403	**GN**	H *GN*	BN
12404	**GN**	H *GN*	BN
12405	**GN**	H *GN*	BN
12406	**GN**	H *GN*	BN
12407	**GN**	H *GN*	BN
12408	**GN**	H *GN*	BN
12409	**GN**	H *GN*	BN
12410	**GN**	H *GN*	BN
12411	**GN**	H *GN*	BN
12412	**GN**	H *GN*	BN
12413	**GN**	H *GN*	BN
12414	**GN**	H *GN*	BN
12415	**GN**	H *GN*	BN
12416	**GN**	H *GN*	BN
12417	**GN**	H *GN*	BN
12418	**GN**	H *GN*	BN
12419	**GN**	H *GN*	BN
12420	**GN**	H *GN*	BN
12421	**GN**	H *GN*	BN
12422	**GN**	H *GN*	BN
12423	**GN**	H *GN*	BN
12424	**GN**	H *GN*	BN
12425	**GN**	H *GN*	BN
12426	**GN**	H *GN*	BN
12427	**GN**	H *GN*	BN
12428	**GN**	H *GN*	BN
12429	**GN**	H *GN*	BN
12430	**GN**	H *GN*	BN
12431	**GN**	H *GN*	BN
12432	**GN**	H *GN*	BN
12433	**GN**	H *GN*	BN
12434	**GN**	H *GN*	BN
12435	**GN**	H *GN*	BN
12436	**GN**	H *GN*	BN
12437	**GN**	H *GN*	BN
12438	**GN**	H *GN*	BN
12439	**GN**	H *GN*	BN
12440	**GN**	H *GN*	BN
12441	**GN**	H *GN*	BN
12442	**GN**	H *GN*	BN
12443	**GN**	H *GN*	BN
12444	**GN**	H *GN*	BN
12445	**GN**	H *GN*	BN
12446	**GN**	H *GN*	BN
12447	**GN**	H *GN*	BN
12448	**GN**	H *GN*	BN
12449	**GN**	H *GN*	BN
12450	**GN**	H *GN*	BN
12451	**GN**	H *GN*	BN
12452	**GN**	H *GN*	BN
12453	**GN**	H *GN*	BN
12454	**GN**	H *GN*	BN
12455	**GN**	H *GN*	BN
12456	**GN**	H *GN*	BN
12457	**GN**	H *GN*	BN
12458	**GN**	H *GN*	BN
12459	**GN**	H *GN*	BN
12460	**GN**	H *GN*	BN
12461	**GN**	H *GN*	BN
12462	**GN**	H *GN*	BN
12463	**GN**	H *GN*	BN
12464	**GN**	H *GN*	BN
12465	**GN**	H *GN*	BN
12466	**GN**	H *GN*	BN
12467	**GN**	H *GN*	BN
12468	**GN**	H *GN*	BN
12469	**GN**	H *GN*	BN
12470	**GN**	H *GN*	BN
12471	**GN**	H *GN*	BN
12472	**GN**	H *GN*	BN
12473	**GN**	H *GN*	BN
12474	**GN**	H *GN*	BN
12475	**GN**	H *GN*	BN
12476	**GN**	H *GN*	BN
12477	**GN**	H *GN*	BN
12478	**GN**	H *GN*	BN
12479	**GN**	H *GN*	BN
12480	**GN**	H *GN*	BN
12481	**GN**	H *GN*	BN

12482 **GN** H *GN* BN
12483 **GN** H *GN* BN
12484 **GN** H *GN* BN
12485 **GN** H *GN* BN
12486 **GN** H *GN* BN
12487 **GN** H *GN* BN
12488 **GN** H *GN* BN
12489 **GN** H *GN* BN
12513 **GN** H *GN* BN
12514 **GN** H *GN* BN
12515 **GN** H *GN* BN
12516 **GN** H *GN* BN
12517 **GN** H *GN* BN
12518 **GN** H *GN* BN
12519 **GN** H *GN* BN
12520 **GN** H *GN* BN
12521 **GN** H *GN* BN
12522 **GN** H *GN* BN
12523 **GN** H *GN* BN
12524 **GN** H *GN* BN
12525 **GN** H *GN* BN
12526 **GN** H *GN* BN
12527 **GN** H *GN* BN
12528 **GN** H *GN* BN
12529 **GN** H *GN* BN
12530 **GN** H *GN* BN
12531 **GN** H *GN* BN
12532 **GN** H *GN* BN
12533 **GN** H *GN* BN
12534 **GN** H *GN* BN
12535 **GN** H *GN* BN
12536 **GN** H *GN* BN
12537 **GN** H *GN* BN
12538 **GN** H *GN* BN

AA11 (FK) CORRIDOR FIRST

Dia. AA101. Mark 1. 42/– 2T. ETH 3.

13225–13230. Lot No. 30381 Swindon 1959. B4 bogies. 33 t.
13318–13341. Lot No. 30667 Swindon 1962. Commonwealth bogies. 36 t.

f Fitted with fluorescent lighting.

13225 k **RR** H EC
13227 xk **CH** RV *ON* CP
13228 xk **M** SP BT
13229 xk **M** SP *ON* BT
13230 xk **M** SP *ON* BT
13318 RS *ON* BN
13321 x **M** WC *ON* CS
13323 xf **M** WC CS
13331 vf **N** LW CP
13341 f **WR** RS *ON* BN

AA1D (FK) CORRIDOR FIRST

Dia. AA109. Mark 2D. Air conditioned (Stones). 13585–13607 require at least 800 V train supply. 42/– 2T. B4 bogies. ETH 5.

Lot No. 30825 Derby 1971–2. 34.5 t.

13575 **N** H OM
13582 E KN
13585 CN KN
13604 CN BN
13607 CN FK

AB11 (BFK) CORRIDOR BRAKE FIRST

Dia. AB101. Mark 1. 24/– 1T. Commonwealth bogies. ETH 2.

17007. Lot No. 30382 Swindon 1959. 35 t.
17013–17019. Lot No. 30668 Swindon 1961. 36 t.
17023. Lot No. 30718 Swindon 1963. Metal window frames. 36 t.

Originally numbered 14007/13/15/18/19/23.

17007 x **PC** MN *OS* SL
17013 **M** FS *OS* SZ

17015 x	**BG**	RS	*ON*	BN	17019 x	**M**	NE	*OS*	BQ
17018 v	**CH**	BR	*ON*	TM	17023 x	**G**	RS	*ON*	BN

AB1Z (BFK) CORRIDOR BRAKE FIRST

Dia. AB102. Mark 2. Pressure ventilated. 24/– 1T. B4 bogies. ETH 4.

Lot No. 30756 Derby 1966. 31.5 t.

Originally numbered 14039.

17039 v	**RX**	H	*E*	CD

AB1A (BFK) CORRIDOR BRAKE FIRST

Dia. AB103. Mark 2A. Pressure ventilated. 24/– 1T. B4 bogies. ETH 4.

17056–17077. Lot No. 30775 Derby 1967–8. 32 t.
17086–17102. Lot No. 30786 Derby 1968. 32 t.

Originally numbered 14056–102. 17090 was numbered 35503 for a time when declassified.

17056	**CH**	RV	*ON*	CP	17090 v	**RR**	BR		TM
17058	**N**	H		KN	17091 v	**RR**	H		LT
17064 v	**RR**	H		LT	17096	**G**	MN		SL
17077	**H**	H	*CA*	CF	17099 v	**RR**	H		LT
17086	**H**	H	*CA*	CF	17102	**M**	WC	*ON*	CS

AB1D (BFK) CORRIDOR BRAKE FIRST

Dia. AB106. Mark 2D. Air conditioned (Stones equipment). 17163–17172 require at least 800 V train supply. 24/– 1T. B4 Bogies. ETH 5.

Lot No. 30823 Derby 1971–2. 33.5 t.

Non-Standard Livery: 17141 & 17164 are as **WV** without lining.

Originally numbered 14141–72.

17141	**0**	CN		FK	17163		H		KN
17144		SO	*SO*	ZA	17164	**0**	RV	*SL*	BY
17146		SO	*SO*	ZA	17165		CN		FK
17148		H		KN	17166		H		LT
17151		VS		CP	17167		VS	*ON*	CP
17153	**WR**	CN		CS	17168	**M**	WC	*ON*	CS
17155		H		KN	17169		CN		CS
17156		CN		DY	17170		CN		DY
17159	**CH**	RV	*SL*	BY	17171		E		KM
17161		E		OM	17172		CN		FK

AE1G (BFO) OPEN BRAKE FIRST

Dia. AE101. Mark 3B. Air conditioned. Fitted with hydraulic handbrake. Refurbished with table lamps and burgundy seat trim. 36/– 1T (w 35/– 1T) BT10 bogies. pg. d. ETH 5X.

Lot No. 30990 Derby 1986. 35.81 t.

```
17173     V  P  VW  PC
17174     V  P  VW  PC
17175 w   V  P  VW  PC
```

AB31 (BCK) CORRIDOR BRAKE COMPOSITE

Dia. AB301 (AB302*). Mark 1. There are two variants depending upon whether the standard class compartments have armrests. Each vehicle has two first class and three standard class compartments. 12/18 2T (12/24 2T *). ETH 2.

21224. Lot No. 30245. Metro-Cammell 1958. B4 bogies. 33 t.
21236–21246. Lot No. 30669 Swindon 1961–2. Commonwealth bogies. 36 t.
21252–21256. Lot No. 30731 Derby 1963. Commonwealth bogies. 37 t.
21266–21272. Lot No. 30732 Derby 1964. Commonwealth bogies. 37 t.

```
21224     RB  VS  ON  CP
21236 v   M   RV  OS  ZG
21241 x   M   SP  ON  BT
21245 x   CC  RS  ON  BN
21246     BG  RS  ON  BN
21252 v   G   MH  ON  RL
21256 x   M   WC  ON  CS
21266 *       FS      SZ
21268 *       FS      SZ
21269 *   WV  RS  ON  BN
21272 x*  CH  RV  ON  CP
```

AA21 (SK) CORRIDOR STANDARD

Dia. AA201 (AA202*). Mark 1. There are two variants depending upon whether the standard class compartments have armrests. Each vehicle has eight compartments. All remaining vehicles have metal window frames and melamine interior panelling. Commonwealth bogies. –/48 2T (–/64 2T *). ETH 4.

25729–25893. Lot No. 30685 Derby 1961–2. 36 t.
25955. Lot No. 30686 Derby 1962. 36 t.
26013. Lot No. 30719 Derby 1962. 37 t.

Non-Standard Livery: 25837 and 25893 are Pilkington's K (green with white red chevron and light blue block).

f Facelifted with fluorescent lighting.
t Rebuilt internally as TSO using components from 4936. -/64 2T.

These coaches were renumbered 18729–19013 for a time.

```
25729 x*f  M   WC  ON  TM
25756 x    M   WC  ON  TM
25767 x    CH  WC  ON  TM
25806 xt   M   WC  ON  CS
25808 x    M   WC  ON  TM
25837 x    0   WC  ON  CS
25862 x    M   WC  ON  TM
25893 x    CH  WC  ON  TM
25955 x*f  M   WC  ON  CS
26013 x    0   WC  ON  CS
```

AB21 (BSK) CORRIDOR BRAKE STANDARD

Dia. AB201 (AB202*). Mark 1. There are two variants depending upon whether the standard class compartments have armrests. Each vehicle has four compartments. Lots 30699, 30721 and 30728 have metal window frames and melamine interior panelling. –/24 1T (–/32 1T*). ETH2.

g Converted to e.t.h. generator vehicle.

34525. Lot No. 30095 Wolverton 1955. Commonwealth bogies. 36 t.
34991. Lot No. 30229 Metro-Cammell 1956–7. Commonwealth bogies. 36 t.
35185–35207. Lot No. 30427 Wolverton 1959. B4 bogies. 33 t.
35317–35333. Lot No. 30699 Wolverton 1962–3. Commonwealth bogies. 37 t.
35407, 35452–35486. Lot No. 30721 Wolverton 1963. Commonwealth Bogies. 37 t.
35449. Lot No. 30728 Wolverton 1963. Commonwealth bogies. 37 t.

Non-Standard Livery: 35407 is in London & North Western Railway livery.

34525	g	**M**	GS		CS
34991	*	**PC**	VS	*ON*	SL
35185	x	**M**	SP		BT
35207	x*	**CC**	VS	*OS*	SL
35317	x	**M**	WT	*ON*	CS
35322	x	**M**	SH	*ON*	CJ
35329	v	**G**	MH	*ON*	RL
35333	x	**CH**	24	*OS*	DI
35407	xg	**0**	SH	*ON*	CJ
35449	x	**CH**	14	*OS*	BQ
35452	x	**RR**	H	*NW*	CP
35453	x	**CH**	RV	*ON*	CP
35457	v	**M**	IE	*OS*	NY
35459	x	**M**	WC	*ON*	CS
35461	x	**CH**	RV		CP
35463	v	**M**	WC	*OS*	CS
35465	x	**WV**	LW	*OS*	CQ
35467	v	**M**	RV	*OS*	KR
35468	v	**M**	NR	*OS*	YM
35469	xg	**CC**	RS	*ON*	BN
35470	v	**CH**	BR	*OS*	TM
35476	v	**CC**	62	*OS*	SK
35479	v	**M**	SV	*OS*	KR
35486	v	**M**	SV	*OS*	KR

AB2A/AB2C (BSK) CORRIDOR BRAKE STANDARD

Dia. AB204. Mark 2A (c 2C). Pressure ventilated. Renumbered from BFK. –/24 1T. B4 bogies. ETH 4.

35507/8/11. Lot No. 30796 Derby 1969–70. 32.5 t.
35510/12–14. Lot No. 30775 Derby 1967–68. 32 t.
35515–18. Lot No. 30786 Derby 1968. 32 t.

* Cage removed from brake compartment.

35507	(14123, 17123)	c	**RR**	H		KN
35508	(14128, 17128)	c	**RR**	CN		CP
35510	(14075, 17075)		**RR**	H		KN
35511	(14130, 17130)	c	**RR**	H		KN
35512	(14057, 17057)	*	**RR**	H		CP
35513	(14063, 17063)	*	**RR**	H	*NW*	CP
35514	(14069, 17069)	*	**RR**	H		CP
35515	(14079, 17079)	*	**RR**	H	*NW*	CP
35516	(14080, 17080)	*	**RR**	H	*NW*	CP
35517	(14088, 17088)	*	**RR**	H	*NW*	CP
35518	(14097, 17097)	*	**RR**	H	*NW*	CP

NAMED COACHES

The following miscellaneous coaches carry names:

1566	CAR 1566
1659	CAMELOT
1683	CAROL
1953	LANCASTRIAN
3065	ORCHID
3066	CHATSWORTH
3068	BEAULIEU
3069	ALNWICK
3105	JULIA
3125	LOCH SHIEL
3130	PAMELA
3174	GLAMIS
3181	MONARCH
3188	SOVEREIGN
3240	PENDENNIS
3267	TREGENNA
3273	RESTORMEL
5132	CLAN MUNRO
5154	CLAN FRASER
5166	CLAN MACKENZIE
5191	CLAN DONALD
5193	CLAN MACLEOD
5212	CAPERKAILZIE
5275	Wendy
5307	Beverley
5350	Dawn
5364	Andrea
5365	Deborah
5373	Felicity
5376	Michaela
5378	Sarah
9385	BALMACARA
9388	BAILECHAUL
9417	Ellen
10569	LEVIATHAN
10729	OLYMPIC
17007	MERCATOR
17077	Catherine
17086	Georgina
17167	ASPINALL
21224	DIRECTORS CAR
35449	ELIZABETH

2. HIGH SPEED TRAIN TRAILER CARS

HSTs normally run in formations of 7 or 8 trailer cars with a Class 43 power car at each end. All trailer cars are classified Mark 3 and have BT10 bogies with disc brakes and central door locking. Heating is by a 415 V three-phase supply and vehicles have air conditioning. Max. Speed is 125 m.p.h.

All vehicles underwent a mid-life refurbishment in the 1980s, and they are at present undergoing a further refurbishment, each train operating company having a different scheme as follows:

First Great Western. Green seat covers and extra partitions between seat bays.
Great North Eastern Railway. New ceiling lighting panels and brown seat covers. First class vehicles have table lamps and imitation walnut plastic end panels.
Virgin Cross-Country. Green seat covers. Standard class vehicles have four seats in the centre of each carriage replaced with a luggage stack.
Virgin West Coast. Red seat covers in first class and light blue seat covers in standard class.
Midland Mainline. Grey seat covers, redesigned seat squabs, side carpeting and two seats in the centre of each carriage replaced with a luggage stack.

All Midland Mainline, First Great Western & GNER and most Virgin Cross-Country vehicles have now been refurbished.

Tops Type Codes

TOPS type codes for HST trailer cars are made up as follows:

(1) Two letters denoting the layout of the vehicle as follows:

GH Open
GJ Open with Guard's compartment.
GK Buffet
GL Kitchen
GN Buffet

(2) A digit for the class of passenger accommodation

1 First
2 Standard (formerly second)
4 Unclassified

(3) A suffix relating to the build of coach.

G Mark 3

Operator Codes

The normal operator codes are given in brackets after the TOPS codes. These are as follows:

TF	Trailer First	TRFK	Trailer Kitchen First
TGS	Trailer Guard's Standard	TRFM	Trailer Modular Buffet First
TRB	Trailer Buffet First	TRSB	Trailer Buffet Standard
TRFB	Trailer Buffet First	TS	Trailer Standard

GN4G (TRB) TRAILER BUFFET FIRST

Dia. GN401. Converted from TRSB by fitting first class seats. Renumbered from 404xx series by subtracting 200. 23/–. p. q.

40204–40228. Lot No. 30883 Derby 1976–7. 36.12 t.
40231. Lot No. 30899 Derby 1978–9. 36.12 t.

40204	**GW**	A	*GW*	PM
40205	**GW**	A	*GW*	PM
40206	**GW**	A	*GW*	PM
40207	**GW**	A	*GW*	PM
40208	**GW**	A	*GW*	PM
40209	**GW**	A	*GW*	PM
40210	**GW**	A	*GW*	PM
40213	**GW**	A	*GW*	PM
40221	**GW**	A	*GW*	PM
40228	**GW**	A	*GW*	PM
40231	**GW**	A	*GW*	PM

GK2G (TRSB) TRAILER BUFFET STANDARD

Dia. GK202. Renumbered from 400xx series by adding 400. –/33 1W. p. q.

40401–40427. Lot No. 30883 Derby 1976–7. 36.12 t.
40429–40437. Lot No. 30899 Derby 1978–9. 36.12 t.

40411/2/32–4 were numbered 40211/2/32–4 for a time when fitted with first class seats.

40401	**V**	P	*VX*	LA
40402	**V**	P	*VX*	LA
40403		P	*VX*	LA
40411	**V**	P	*VX*	LA
40412	**V**	P	*VX*	LA
40414	**V**	P	*VX*	LA
40415	**V**	P	*VX*	LA
40416		P	*VX*	LA
40417		P	*VX*	LA
40418	**V**	P	*VX*	LA
40419	**V**	P	*VX*	LA
40420	**V**	P	*VX*	LA
40422	**V**	P	*VX*	LA
40423	**V**	P	*VX*	LA
40424		P	*VX*	LA
40425		P	*VX*	LA
40426		P	*VX*	LA
40427	**V**	P	*VX*	LA
40429	**V**	P	*VX*	LA
40430	**V**	P	*VX*	LA
40432	**V**	P	*VX*	LA
40433	**V**	P	*VX*	LA
40434	**V**	P	*VX*	LA
40435	**V**	P	*VX*	LA
40436	**V**	P	*VX*	LA
40437	**V**	P	*VX*	LA

GL1G (TRFK) TRAILER KITCHEN FIRST

Dia. GL101. Reclassified from TRUK. p. q. 24/–.

Lot No. 30884 Derby 1976–7. 37 t.

40501		P		ZD
40513		P		ZD

GK1G (TRFM) TRAILER MODULAR BUFFET FIRST

Dia. GK102. Converted to modular catering from TRFB 40719. 17/– 1T. p. q.

Lot No. 30921 Derby 1978–9. 38.16 t.

40619		P	*VX*	LA

GK1G (TRFB) TRAILER BUFFET FIRST

Dia. GK101. These vehicles have larger kitchens than the 402xx and 404xx series vehicles, and are used in trains where full meal service is required. They were renumbered from the 403xx series (in which the seats were unclassified) by adding 400 to previous number. 17/–. p. q.

40700–40721. Lot No. 30921 Derby 1978–9. 38.16 t.
40722–40735. Lot No. 30940 Derby 1979–80. 38.16 t.
40736–40753. Lot No. 30948 Derby 1980–1. 38.16 t.
40754–40757. Lot No. 30966 Derby 1982. 38.16 t.

40700	**MM**	P	*MM*	NL
40701	**MM**	P	*MM*	NL
40702	**MM**	P	*MM*	NL
40703	**GW**	A	*GW*	LA
40704	**GN**	A	*GN*	EC
40705	**GN**	A	*GN*	EC
40706	**GN**	A	*GN*	EC
40707	**GW**	A	*GW*	LA
40708	**MM**	P	*MM*	NL
40709	**GW**	A	*GW*	LA
40710	**GW**	A	*GW*	LA
40711	**GN**	A	*GN*	EC
40712	**GW**	A	*GW*	LA
40713	**GW**	A	*GW*	LA
40714	**GW**	A	*GW*	PM
40715	**GW**	A	*GW*	PM
40716	**GW**	A	*GW*	PM
40717	**GW**	A	*GW*	PM
40718	**GW**	A	*GW*	LA
40720	**GN**	A	*GN*	EC
40721	**GW**	A	*GW*	LA
40722	**GW**	A	*GW*	LA
40723	**V**	A	*VW*	LA
40724	**GW**	A	*GW*	PM
40725	**GW**	A	*GW*	LA
40726	**GW**	A	*GW*	LA
40727	**GW**	A	*GW*	LA
40728	**MM**	P	*MM*	NL
40729	**MM**	P	*MM*	NL
40730	**MM**	P	*MM*	NL
40731	**GW**	A	*GW*	LA
40732	**V**	A	*VW*	LA
40733	**GW**	A	*GW*	LA
40734	**GW**	A	*GW*	PM
40735	**GN**	A	*GN*	EC
40736	**GW**	A	*GW*	LA
40737	**GN**	A	*GN*	EC
40738	**GW**	A	*GW*	LA
40739	**GW**	A	*GW*	PM
40740	**GN**	A	*GN*	EC
40741	**MM**	P	*MM*	NL
40742	**V**	A	*VW*	LA
40743	**GW**	A	*GW*	LA
40744	**GW**	A	*GW*	PM
40745	**GW**	A	*GW*	LA
40746	**MM**	P	*MM*	NL
40747	**GW**	A	*GW*	PM
40748	**GN**	A	*GN*	EC
40749	**MM**	P	*MM*	NL
40750	**GN**	A	*GN*	EC
40751	**MM**	P	*MM*	NL
40752	**GW**	A	*GW*	LA
40753	**MM**	P	*MM*	NL
40754	**MM**	P	*MM*	NL
40755	**GW**	A	*GW*	LA
40756	**MM**	P	*MM*	NL
40757	**GW**	A	*GW*	LA

GH1G (TF) TRAILER FIRST

Dia. GH102. 48/– 2T (w 47/– 2T 1W).

41003–41056. Lot No. 30881 Derby 1976–7. 33.66 t.
41057–41120. Lot No. 30896 Derby 1977–8. 33.66 t.
41121–41148. Lot No. 30938 Derby 1979–80. 33.66 t.
41149–41166. Lot No. 30947 Derby 1980. 33.66 t.

41167–41169. Lot No. 30963 Derby 1982. 33.66 t.
41170. Lot No. 30967 Derby 1982. Former prototype vehicle. 33.66 t.
41179/80. Lot No. 30884 Derby 1976–7. 33.60 t.

s Fitted with centre luggage stack. 46/– 2T 1TD 1W.

41170 was converted from 41001. 41179/80 have been converted from 40505 and 40511 respectively.

41003	p	**GW**	A	*GW*	LA
41004		**GW**	A	*GW*	PM
41005	p	**GW**	A	*GW*	PM
41006		**GW**	A	*GW*	PM
41007	p	**GW**	A	*GW*	PM
41008		**GW**	A	*GW*	PM
41009	p	**GW**	A	*GW*	PM
41010		**GW**	A	*GW*	PM
41011	p	**GW**	A	*GW*	PM
41012		**GW**	A	*GW*	PM
41013	p	**GW**	A	*GW*	PM
41014		**GW**	A	*GW*	PM
41015	p	**GW**	A	*GW*	PM
41016		**GW**	A	*GW*	PM
41017	p	**GW**	A	*GW*	PM
41018		**GW**	A	*GW*	PM
41019	p	**GW**	A	*GW*	PM
41020		**GW**	A	*GW*	PM
41021	p	**GW**	A	*GW*	PM
41022		**GW**	A	*GW*	PM
41023	p	**GW**	A	*GW*	LA
41024		**GW**	A	*GW*	LA
41025	p	**V**	A	*VW*	LA
41026		**V**	A	*VW*	LA
41027	p	**GW**	A	*GW*	LA
41028		**GW**	A	*GW*	LA
41029	p	**GW**	A	*GW*	LA
41030		**GW**	A	*GW*	LA
41031	p	**GW**	A	*GW*	LA
41032		**GW**	A	*GW*	LA
41033	p	**GW**	A	*GW*	LA
41034		**GW**	A	*GW*	LA
41035	p	**V**	A	*VW*	LA
41036	w	**V**	A	*VW*	LA
41037	p	**GW**	A	*GW*	LA
41038		**GW**	A	*GW*	LA
41039		**GN**	A	*GN*	EC
41040		**GN**	A	*GN*	EC
41041	ps	**MM**	P	*MM*	NL
41042		**GW**	A	*GW*	PM
41043	w	**GN**	A	*GN*	EC
41044		**GN**	A	*GN*	EC
41045	w	**V**	P	*VX*	LA
41046	s	**MM**	P	*MM*	NL
41051		**GW**	A	*GW*	LA
41052		**GW**	A	*GW*	LA
41055		**GW**	A	*GW*	LA
41056		**GW**	A	*GW*	LA
41057		**MM**	P	*MM*	NL
41058	s	**MM**	P	*MM*	NL
41059	w	**V**	P	*VX*	LA
41060		**GW**	A	*GW*	LA
41061		**MM**	P	*MM*	NL
41062	w	**MM**	P	*MM*	NL
41063		**MM**	P	*MM*	NL
41064	s	**MM**	P	*MM*	NL
41065		**GW**	A	*GW*	LA
41066	p	**V**	A	*VW*	LA
41067	s	**MM**	P	*MM*	NL
41068	s	**MM**	P	*MM*	NL
41069	s	**MM**	P	*MM*	NL
41070	s	**MM**	P	*MM*	NL
41071		**MM**	P	*MM*	NL
41072	s	**MM**	P	*MM*	NL
41075		**MM**	P	*MM*	NL
41076	s	**MM**	P	*MM*	NL
41077		**MM**	P	*MM*	NL
41078		**MM**	P	*MM*	NL
41079		**MM**	P	*MM*	NL
41080	s	**MM**	P	*MM*	NL
41081	w	**V**	P	*VX*	LA
41082	w		P	*VX*	LA
41083		**MM**	P	*MM*	NL
41084	s	**MM**	P	*MM*	NL
41085	w	**V**	P	*VX*	LA
41086	w	**V**	P	*VX*	LA
41087		**GN**	A	*GN*	EC
41088	w	**GN**	A	*GN*	EC
41089		**GW**	A	*GW*	LA
41090	w	**GN**	A	*GN*	EC
41091		**GN**	A	*GN*	EC
41092	w	**GN**	A	*GN*	EC
41093		**GW**	A	*GW*	LA
41094		**GW**	A	*GW*	LA
41095	w	**V**	P	*VX*	LA
41096	w	**V**	P	*VX*	LA

41097 w	**GN**	A	*GN*	EC
41098 w	**GN**	A	*GN*	EC
41099	**GN**	A	*GN*	EC
41100 w	**GN**	A	*GN*	EC
41101	**GW**	A	*GW*	LA
41102	**GW**	A	*GW*	LA
41103	**GW**	A	*GW*	LA
41104	**GW**	A	*GW*	LA
41105	**GW**	A	*GW*	PM
41106	**GW**	A	*GW*	PM
41107 w	**V**	P	*VX*	LA
41108 w	**V**	P	*VX*	LA
41109 w	**V**	P	*VX*	LA
41110	**GW**	A	*GW*	PM
41111	**MM**	P	*MM*	NL
41112	**MM**	P	*MM*	NL
41113 s	**MM**	P	*MM*	NL
41114 w		P	*VX*	LA
41115		P	*VX*	LA
41116	**GW**	A	*GW*	LA
41117	**MM**	P	*MM*	NL
41118 w	**GN**	A	*GN*	EC
41119 w	**V**	P	*VX*	LA
41120	**GN**	A	*GN*	EC
41121 p	**GW**	A	*GW*	LA
41122	**GW**	A	*GW*	LA
41123 p	**GW**	A	*GW*	PM
41124	**GW**	A	*GW*	PM
41125	**GW**	A	*GW*	PM
41126 p	**GW**	A	*GW*	PM
41127 p	**GW**	A	*GW*	PM
41128	**GW**	A	*GW*	PM
41129 p	**GW**	A	*GW*	PM
41130	**GW**	A	*GW*	PM
41131 p	**GW**	A	*GW*	LA
41132	**GW**	A	*GW*	LA
41133 p	**GW**	A	*GW*	LA
41134	**GW**	A	*GW*	LA
41135 p	**GW**	A	*GW*	LA
41136	**GW**	A	*GW*	LA
41137 p	**GW**	A	*GW*	PM
41138	**GW**	A	*GW*	PM
41139 p	**GW**	A	*GW*	LA
41140	**GW**	A	*GW*	LA
41141 p	**GW**	A	*GW*	PM
41142	**GW**	A	*GW*	PM
41143 p	**GW**	A	*GW*	LA
41144	**GW**	A	*GW*	LA
41145 p	**GW**	A	*GW*	PM
41146	**GW**	A	*GW*	PM
41147 w	**V**	P	*VX*	LA
41148 w	**V**	P	*VX*	LA
41149 w	**V**	P	*VX*	LA
41150 w	**GN**	A	*GN*	EC
41151	**GN**	A	*GN*	EC
41152	**GN**	A	*GN*	EC
41153	**MM**	P	*MM*	NL
41154 s	**MM**	P	*MM*	NL
41155	**MM**	P	*MM*	NL
41156	**MM**	P	*MM*	NL
41157	**GW**	A	*GW*	LA
41158	**GW**	A	*GW*	LA
41159 w	**V**	P	*VX*	LA
41160 w	**V**	P	*VX*	LA
41161 w	**V**	P	*VX*	LA
41162 w		P	*VX*	LA
41163 w		P	*VX*	LA
41164 p	**V**	A	*VW*	LA
41165 w		P	*VX*	LA
41166 w		P	*VX*	LA
41167 w	**V**	P	*VX*	LA
41168 w	**V**	P	*VX*	LA
41169 w	**V**	P	*VX*	LA
41170	**GN**	A	*GN*	EC
41179	**GW**	A	*GW*	PM
41180	**GW**	A	*GW*	PM

GH2G (TS) TRAILER STANDARD

Dia. GH203. –/76 2T.

42003–42090/42362. Lot No. 30882 Derby 1976–7. 33.60 t.
42091–42250. Lot No. 30897 Derby 1977–9. 33.60 t.
42251–42305. Lot No. 30939 Derby 1979–80. 33.60 t.
42306–42322. Lot No. 30969 Derby 1982. 33.60 t.
42323–42341. Lot No. 30983 Derby 1984–5. 33.60 t.
42342/60. Lot No. 30949 Derby 1982. 33.47 t. Converted from TGS.
42343/5. Lot No. 30970 Derby 1982. 33.47 t. Converted from TGS.
42344/61. Lot No. 30964 Derby 1982. 33.47 t. Converted from TGS.

42346/7/50/1. Lot No. 30881 Derby 1976–7. 33.66 t. Converted from TF.
42348/9. Lot No. 30896 Derby 1977–8. 33.66 t. Converted from TF.
42353/5–7. Lot No. 30967 Derby 1982. Ex prototype vehicles. 33.66 t.
42352/4. Lot No. 30897 Derby 1977. Were TF from 1983 to 1992. 33.66 t.

s Centre luggage stack –/72 2T.
t Centre luggage stack –/72 2T. Fitted with pt.
u Centre luggage stack –/74 2T (w –72 2T 1W).
* disabled persons toilet and 5 tip-up seats. –/65 1T 1TD.
§ –/70 2T 2W.

42158 was also numbered 41177 for a time when fitted with first class seats.

42003		**GW**	A	*GW*	PM
42004	*	**GW**	A	*GW*	LA
42005		**GW**	A	*GW*	PM
42006		**GW**	A	*GW*	PM
42007	*	**GW**	A	*GW*	LA
42008	*	**GW**	A	*GW*	PM
42009		**GW**	A	*GW*	PM
42010		**GW**	A	*GW*	PM
42012	*	**GW**	A	*GW*	PM
42013		**GW**	A	*GW*	PM
42014		**GW**	A	*GW*	PM
42015	*	**GW**	A	*GW*	PM
42016		**GW**	A	*GW*	PM
42017		**GW**	A	*GW*	PM
42018	*	**GW**	A	*GW*	PM
42019		**GW**	A	*GW*	PM
42020		**GW**	A	*GW*	PM
42021	*	**GW**	A	*GW*	PM
42022		**GW**	A	*GW*	PM
42023		**GW**	A	*GW*	PM
42024	*	**GW**	A	*GW*	PM
42025		**GW**	A	*GW*	PM
42026		**GW**	A	*GW*	PM
42027		**GW**	A	*GW*	PM
42028		**GW**	A	*GW*	PM
42029		**GW**	A	*GW*	PM
42030	*	**GW**	A	*GW*	PM
42031		**GW**	A	*GW*	PM
42032		**GW**	A	*GW*	PM
42033		**GW**	A	*GW*	LA
42034		**GW**	A	*GW*	LA
42035		**GW**	A	*GW*	LA
42036		**V**	A	*VW*	LA
42037		**V**	A	*VW*	LA
42038		**V**	A	*VW*	LA
42039		**GW**	A	*GW*	LA
42040		**GW**	A	*GW*	LA
42041		**GW**	A	*GW*	LA
42042		**GW**	A	*GW*	LA
42043		**GW**	A	*GW*	LA
42044		**GW**	A	*GW*	LA
42045		**GW**	A	*GW*	LA
42046		**GW**	A	*GW*	LA
42047		**GW**	A	*GW*	LA
42048		**GW**	A	*GW*	LA
42049		**GW**	A	*GW*	LA
42050		**GW**	A	*GW*	LA
42051	§	**V**	A	*VW*	LA
42052		**V**	A	*VW*	LA
42053		**V**	A	*VW*	LA
42054		**GW**	A	*GW*	PM
42055		**GW**	A	*GW*	LA
42056		**GW**	A	*GW*	LA
42057		**GN**	A	*GN*	EC
42058		**GN**	A	*GN*	EC
42059		**GN**	A	*GN*	EC
42060		**GW**	A	*GW*	PM
42061		**GW**	A	*GW*	PM
42062	*	**GW**	A	*GW*	LA
42063		**GN**	A	*GN*	EC
42064		**GN**	A	*GN*	EC
42065		**GN**	A	*GN*	EC
42066	*	**GW**	A	*GW*	LA
42067		**GW**	A	*GW*	LA
42068		**GW**	A	*GW*	LA
42069	*	**GW**	A	*GW*	PM
42070		**GW**	A	*GW*	PM
42071		**GW**	A	*GW*	PM
42072		**GW**	A	*GW*	PM
42073		**GW**	A	*GW*	PM
42074		**GW**	A	*GW*	PM
42075		**GW**	A	*GW*	PM
42076		**GW**	A	*GW*	LA
42077		**GW**	A	*GW*	LA
42078		**GW**	A	*GW*	LA
42079		**GW**	A	*GW*	PM
42080		**GW**	A	*GW*	PM
42081	*	**GW**	A	*GW*	LA

42082	*	**GW**	A	*GW*	PM
42083		**GW**	A	*GW*	LA
42084	s	**V**	P	*VX*	LA
42085	t	**V**	P	*VX*	LA
42086	s	**V**	P	*VX*	LA
42087	s	**V**	P	*VX*	LA
42088			P	*VX*	LA
42089		**GW**	A	*GW*	PM
42090			P	*VX*	LA
42091			P	*VX*	LA
42092			P	*VX*	LA
42093			P	*VX*	LA
42094			P	*VX*	LA
42095			P	*VX*	LA
42096		**GW**	A	*GW*	LA
42097	§	**V**	A	*VW*	LA
42098		**GW**	A	*GW*	PM
42099		**GW**	A	*GW*	LA
42100	u	**MM**	P	*MM*	NL
42101	uw	**MM**	P	*MM*	NL
42102	u	**MM**	P	*MM*	NL
42103	s	**V**	P	*VX*	LA
42104		**GN**	A	*GN*	EC
42105	s	**V**	P	*VX*	LA
42106		**GN**	A	*GN*	EC
42107		**GW**	A	*GW*	LA
42108			P	*VX*	LA
42109			P	*VX*	LA
42110			P	*VX*	LA
42111	u	**MM**	P	*MM*	NL
42112	u	**MM**	P	*MM*	NL
42113	u	**MM**	P	*MM*	NL
42115	s	**V**	P	*VX*	LA
42116	s	**V**	P	*VX*	LA
42117	s	**V**	P	*VX*	LA
42118		**GW**	A	*GW*	PM
42119	u	**MM**	P	*MM*	NL
42120	u	**MM**	P	*MM*	NL
42121	u	**MM**	P	*MM*	NL
42122		**V**	A	*VW*	LA
42123	u	**MM**	P	*MM*	NL
42124	u	**MM**	P	*MM*	NL
42125	u	**MM**	P	*MM*	NL
42126		**GW**	A	*GW*	LA
42127	s	**V**	P	*VX*	LA
42128	s	**V**	P	*VX*	LA
42129		**GW**	A	*GW*	LA
42130			P	*VX*	LA
42131	u	**MM**	P	*MM*	NL
42132	u	**MM**	P	*MM*	NL
42133	u	**MM**	P	*MM*	NL
42134		**V**	A	*VW*	LA
42135	u	**MM**	P	*MM*	NL
42136	u	**MM**	P	*MM*	NL
42137	u	**MM**	P	*MM*	NL
42138	*	**GW**	A	*GW*	PM
42139	u	**MM**	P	*MM*	NL
42140	u	**MM**	P	*MM*	NL
42141	u	**MM**	P	*MM*	NL
42143		**GW**	A	*GW*	LA
42144		**GW**	A	*GW*	LA
42145		**GW**	A	*GW*	LA
42146		**GN**	A	*GN*	EC
42147	u	**MM**	P	*MM*	NL
42148	u	**MM**	P	*MM*	NL
42149	u	**MM**	P	*MM*	NL
42150		**GN**	A	*GN*	EC
42151	uw	**MM**	P	*MM*	NL
42152	u	**MM**	P	*MM*	NL
42153	u	**MM**	P	*MM*	NL
42154			A	*GN*	EC
42155	uw	**MM**	P	*MM*	NL
42156	u	**MM**	P	*MM*	NL
42157	u	**MM**	P	*MM*	NL
42158		**GN**	A	*GN*	EC
42159			P	*VX*	LA
42160			P	*VX*	LA
42161			P	*VX*	LA
42162	s	**V**	P	*VX*	LA
42163	uw	**MM**	P	*MM*	NL
42164	u	**MM**	P	*MM*	NL
42165	u	**MM**	P	*MM*	NL
42166	t	**V**	P	*VX*	LA
42167	s	**V**	P	*VX*	LA
42168	t	**V**	P	*VX*	LA
42169	s	**V**	P	*VX*	LA
42170	s	**V**	P	*VX*	LA
42171		**GN**	A	*GN*	EC
42172		**GN**	A	*GN*	EC
42173	s	**V**	P	*VX*	LA
42174	s	**V**	P	*VX*	LA
42175	s	**V**	P	*VX*	LA
42176	t	**V**	P	*VX*	LA
42177	s	**V**	P	*VX*	LA
42178	s	**V**	P	*VX*	LA
42179		**GN**	A	*GN*	EC
42180		**GN**	A	*GN*	EC
42181		**GN**	A	*GN*	EC
42182		**GN**	A	*GN*	EC
42183	*	**GW**	A	*GW*	LA
42184		**GW**	A	*GW*	LA
42185		**GW**	A	*GW*	LA

42186	**GN**	A	*GN*	EC
42187 t	**V**	P	*VX*	LA
42188 s	**V**	P	*VX*	LA
42189 s	**V**	P	*VX*	LA
42190	**GN**	A	*GN*	EC
42191	**GN**	A	*GN*	EC
42192	**GN**	A	*GN*	EC
42193	**GN**	A	*GN*	EC
42194 uw	**MM**	P	*MM*	NL
42195 s	**V**	P	*VX*	LA
42196	**GW**	A	*GW*	LA
42197	**GW**	A	*GW*	PM
42198	**GN**	A	*GN*	EC
42199	**GN**	A	*GN*	EC
42200 *	**GW**	A	*GW*	LA
42201 *	**GW**	A	*GW*	LA
42202 *	**GW**	A	*GW*	LA
42203	**GW**	A	*GW*	LA
42204	**GW**	A	*GW*	LA
42205 u	**MM**	P	*MM*	NL
42206 *	**GW**	A	*GW*	LA
42207 *	**GW**	A	*GW*	LA
42208	**GW**	A	*GW*	LA
42209	**GW**	A	*GW*	LA
42210 u	**MM**	P	*MM*	NL
42211 *	**GW**	A	*GW*	PM
42212	**GW**	A	*GW*	PM
42213	**GW**	A	*GW*	PM
42214	**GW**	A	*GW*	PM
42215	**GN**	A	*GN*	EC
42216	**GW**	A	*GW*	LA
42217 t	**V**	P	*VX*	LA
42218 t	**V**	P	*VX*	LA
42219	**GN**	A	*GN*	EC
42220 uw	**MM**	P	*MM*	NL
42221	**GW**	A	*GW*	LA
42222 t	**V**	P	*VX*	LA
42223 s	**V**	P	*VX*	LA
42224 s	**V**	P	*VX*	LA
42225 u	**MM**	P	*MM*	NL
42226	**GN**	A	*GN*	EC
42227 u	**MM**	P	*MM*	NL
42228 u	**MM**	P	*MM*	NL
42229 u	**MM**	P	*MM*	NL
42230 u	**MM**	P	*MM*	NL
42231		P	*VX*	LA
42232		P	*VX*	LA
42233		P	*VX*	LA
42234		P	*VX*	LA
42235	**GN**	A	*GN*	EC
42236	**GW**	A	*GW*	PM
42237	**V**	P	*VX*	LA
42238	**V**	P	*VX*	LA
42239	**V**	P	*VX*	LA
42240	**GN**	A	*GN*	EC
42241	**GN**	A	*GN*	EC
42242	**GN**	A	*GN*	EC
42243	**GN**	A	*GN*	EC
42244	**GN**	A	*GN*	EC
42245	**GW**	A	*GW*	LA
42246 s	**V**	P	*VX*	LA
42247 t	**V**	P	*VX*	LA
42248 s	**V**	P	*VX*	LA
42249 s	**V**	P	*VX*	LA
42250	**GW**	A	*GW*	LA
42251 *	**GW**	A	*GW*	PM
42252	**GW**	A	*GW*	LA
42253	**GW**	A	*GW*	LA
42254 s	**V**	P	*VX*	LA
42255 *	**GW**	A	*GW*	PM
42256	**GW**	A	*GW*	PM
42257	**GW**	A	*GW*	PM
42258 s	**V**	P	*VX*	LA
42259 *	**GW**	A	*GW*	PM
42260	**GW**	A	*GW*	PM
42261	**GW**	A	*GW*	PM
42262 s	**V**	P	*VX*	LA
42263	**GW**	A	*GW*	PM
42264 *	**GW**	A	*GW*	PM
42265	**GW**	A	*GW*	LA
42266 s	**V**	P	*VX*	LA
42267 *	**GW**	A	*GW*	PM
42268 t	**GW**	A	*GW*	LA
42269	**GW**	A	*GW*	PM
42270 s	**V**	P	*VX*	LA
42271 *	**GW**	A	*GW*	LA
42272	**GW**	A	*GW*	LA
42273	**GW**	A	*GW*	LA
42274 t	**V**	P	*VX*	LA
42275 *	**GW**	A	*GW*	LA
42276	**GW**	A	*GW*	LA
42277	**GW**	A	*GW*	LA
42278 s	**V**	P	*VX*	LA
42279 *	**GW**	A	*GW*	LA
42280	**GW**	A	*GW*	LA
42281	**GW**	A	*GW*	LA
42282 s	**V**	P	*VX*	LA
42283	**GW**	A	*GW*	LA
42284	**GW**	A	*GW*	PM
42285	**GW**	A	*GW*	PM
42286 s	**V**	P	*VX*	LA
42287 *	**GW**	A	*GW*	LA

42288		**GW**	A	*GW*	LA
42289		**GW**	A	*GW*	LA
42290	t	**V**	P	*VX*	LA
42291	*	**GW**	A	*GW*	PM
42292	*	**GW**	A	*GW*	LA
42293		**GW**	A	*GW*	PM
42294	s	**V**	P	*VX*	LA
42295	*	**GW**	A	*GW*	LA
42296		**GW**	A	*GW*	LA
42297		**GW**	A	*GW*	LA
42298	s	**V**	P	*VX*	LA
42299	*	**GW**	A	*GW*	PM
42300		**GW**	A	*GW*	PM
42301		**GW**	A	*GW*	PM
42302	s	**V**	P	*VX*	LA
42303	t	**V**	P	*VX*	LA
42304	s	**V**	P	*VX*	LA
42305	s	**V**	P	*VX*	LA
42306			P	*VX*	LA
42307			P	*VX*	LA
42308			P	*VX*	LA
42309			P	*VX*	LA
42310	s	**V**	P	*VX*	LA
42311	t	**V**	P	*VX*	LA
42312	s	**V**	P	*VX*	LA
42313	s	**V**	P	*VX*	LA
42314	s	**V**	P	*VX*	LA
42315	t	**V**	P	*VX*	LA
42316	s	**V**	P	*VX*	LA
42317	s	**V**	P	*VX*	LA
42318	s	**V**	P	*VX*	LA
42319	t	**V**	P	*VX*	LA
42320	s	**V**	P	*VX*	LA
42321	s	**V**	P	*VX*	LA
42322			P	*VX*	LA
42323		**GN**	A	*GN*	EC
42324	uw	**MM**	P	*MM*	NL
42325		**GW**	A	*GW*	PM
42326	s	**V**	P	*VX*	LA
42327	uw	**MM**	P	*MM*	NL
42328	uw	**MM**	P	*MM*	NL
42329	uw	**MM**	P	*MM*	NL
42330		**V**	P	*VX*	LA
42331	uw	**MM**	P	*MM*	NL
42332		**GW**	A	*GW*	PM
42333		**GW**	A	*GW*	LA
42334	s	**V**	P	*VX*	LA
42335	uw	**MM**	P	*MM*	NL
42336	s	**V**	P	*VX*	LA
42337	uw	**MM**	P	*MM*	NL
42338	s	**V**	P	*VX*	LA
42339	uw	**MM**	P	*MM*	NL
42340		**GN**	A	*GN*	EC
42341	uw	**MM**	P	*MM*	NL

42342	(44082)		**V**	A	*VW*	LA
42343	(44095)		**GW**	A	*GW*	LA
42344	(44092)	*	**GW**	A	*GW*	PM
42345	(44096)	*	**GW**	A	*GW*	LA
42346	(41053)		**GW**	A	*GW*	PM
42347	(41054)	*	**GW**	A	*GW*	PM
42348	(41073)	*	**GW**	A	*GW*	LA
42349	(41074)		**GW**	A	*GW*	PM
42350	(41047)		**GW**	A	*GW*	LA
42351	(41048)		**GW**	A	*GW*	PM
42352	(42142, 41176)	u	**MM**	P	*MM*	NL
42353	(42001, 41171)	s	**V**	P	*VX*	LA
42354	(42114, 41175)		**GN**	A	*GN*	EC
42355	(42000, 41172)	§	**V**	A	*VW*	LA
42356	(42002, 41173)		**GW**	A	*GW*	PM
42357	(41002, 41174)		**V**	A	*VW*	LA
42360	(44084, 45084)		**GW**	A	*VW*	PM
42361	(44099)		**GW**	A	*GW*	PM
42362	(42011, 41178)		**GW**	A	*GW*	PM

GJ2G (TGS) TRAILER GUARD'S STANDARD

Dia. GJ205. –/65 1T (w –/63 1T 1W). pg.

44000. Lot No. 30953 Derby 1980. 33.47 t.
44001–44090. Lot No. 30949 Derby 1980–2. 33.47 t.
44091–44094. Lot No. 30964 Derby 1982. 33.47 t.
44097–44101. Lot No. 30970 Derby 1982. 33.47 t.

s Fitted with centre luggage stack –/63 1T.
t Fitted with centre luggage stack –/61 1T.

44000	t	**V**	P	*VX*	LA
44001	w	**GW**	A	*GW*	LA
44002	w	**GW**	A	*GW*	PM
44003	w	**GW**	A	*GW*	PM
44004	w	**GW**	A	*GW*	PM
44005	w	**GW**	A	*GW*	PM
44006	w	**GW**	A	*GW*	PM
44007	w	**GW**	A	*GW*	PM
44008	w	**GW**	A	*GW*	PM
44009	w	**GW**	A	*GW*	PM
44010	w	**GW**	A	*GW*	PM
44011	w	**GW**	A	*GW*	LA
44012		**V**	A	*VW*	LA
44013	w	**GW**	A	*GW*	LA
44014	w	**GW**	A	*GW*	LA
44015	w	**GW**	A	*GW*	LA
44016	w	**GW**	A	*GW*	LA
44017		**V**	A	*VW*	LA
44018	w	**GW**	A	*GW*	LA
44019	w	**GN**	A	*GN*	EC
44020	w	**GW**	A	*GW*	PM
44021	t	**V**	P	*VX*	LA
44022	w	**GW**	A	*GW*	LA
44023	w	**GW**	A	*GW*	PM
44024	w	**GW**	A	*GW*	PM
44025	w	**GW**	A	*GW*	LA
44026	w	**GW**	A	*GW*	PM
44027	s	**MM**	P	*MM*	NL
44028	w	**GW**	A	*GW*	LA
44029	w	**GW**	A	*GW*	PM
44030	w	**GW**	A	*GW*	PM
44031		**V**	A	*VW*	LA
44032	w	**GW**	A	*GW*	PM
44033	w	**GW**	A	*GW*	LA
44034	w	**GW**	A	*GW*	LA
44035	w	**GW**	A	*GW*	LA
44036	w	**GW**	A	*GW*	PM
44037	w	**GW**	A	*GW*	LA
44038	w	**GW**	A	*GW*	PM
44039	w	**GW**	A	*GW*	LA
44040	w	**GW**	A	*GW*	PM
44041	s	**MM**	P	*MM*	NL
44042	t	**V**	P	*VX*	LA
44043	w	**GW**	A	*GW*	LA
44044	s	**MM**	P	*MM*	NL
44045	w	**GN**	A	*GN*	EC
44046	s	**MM**	P	*MM*	NL
44047	s	**MM**	P	*MM*	NL
44048	s	**MM**	P	*MM*	NL
44049	w	**GW**	A	*GW*	LA
44050	s	**MM**	P	*MM*	NL
44051	s	**MM**	P	*MM*	NL
44052	s	**MM**	P	*MM*	NL
44053	w		P	*VX*	LA
44054	s	**MM**	P	*MM*	NL
44055	t	**V**	P	*VX*	LA
44056	w	**GN**	A	*GN*	EC
44057	t	**V**	P	*VX*	LA
44058	w	**GN**	A	*GN*	EC
44059	w	**GW**	A	*GW*	LA
44060		**V**	P	*VX*	LA
44061	w	**GN**	A	*GN*	EC
44062	t	**V**	P	*VX*	LA
44063	w	**GN**	A	*GN*	EC
44064	w	**GW**	A	*GW*	LA
44065	t	**V**	P	*VX*	LA
44066	w	**GW**	A	*GW*	LA
44067	w	**GW**	A	*GW*	PM
44068	t	**V**	P	*VX*	LA
44069	t	**V**	P	*VX*	LA
44070	s	**MM**	P	*MM*	NL
44071	s	**MM**	P	*MM*	NL
44072			P	*VX*	LA
44073	s	**MM**	P	*MM*	NL
44074			P	*VX*	LA
44075	t	**V**	P	*VX*	LA
44076			P	*VX*	LA
44077	w	**GN**	A	*GN*	EC

44078	t	**V**	P	*VX*	LA
44079	t	**V**	P	*VX*	LA
44080	w	**GN**	A	*GN*	EC
44081	t	**V**	P	*VX*	LA
44083	s	**MM**	P	*MM*	NL
44085	s	**MM**	P	*MM*	NL
44086	w	**GW**	A	*GW*	LA
44087			P	*VX*	LA
44088			P	*VX*	LA
44089	t	**V**	P	*VX*	LA
44090	t	**V**	P	*VX*	LA
44091	t	**V**	P	*VX*	LA
44093	w	**GW**	A	*GW*	LA
44094	w	**GN**	A	*GN*	EC
44097	t	**V**	P	*VX*	LA
44098	w	**GN**	A	*GN*	EC
44100	t	**V**	P	*VX*	LA
44101	t	**V**	P	*VX*	LA

3. NON-PASSENGER-CARRYING COACHING STOCK

The notes shown for locomotive-hauled passenger stock generally apply also to non-passenger-carrying coaching stock (often abbreviated to NPCCS).

TOPS TYPE CODES

TOPS type codes for NPCCS are made up as follows:

(1) Two letters denoting the type of the vehicle:

AX	Nightstar generator van
AY	Eurostar barrier vehicle
NA	Propelling control vehicle.
NB	High security brake van (100 m.p.h.).
NC	Gangwayed brake van modified for newspaper conveyance (100 m.p.h.).
ND	Gangwayed brake van (90 m.p.h.).
NE	Gangwayed brake van (100 m.p.h.).
NF	Gangwayed brake van with guard's safety equipment removed.
NG	Motorail loading wagon.
NH	Gangwayed brake van (110 m.p.h.).
NI	High security brake van (110 m.p.h.).
NJ	General utility van (90 m.p.h.).
NK	High security general utility van (100 m.p.h.).
NL	Newspaper van.
NN	Courier vehicle.
NO	General utility van (100 m.p.h. e.t.h. wired).
NP	General utility van for Post Office use or Motorail van (110 m.p.h.).
NR	BAA container van (100 m.p.h.).
NS	Post office sorting van.
NT	Post office stowage van.
NU	Brake post office stowage van.
NV	Motorail van (side loading).
NX	Motorail van (100 m.p.h.).
NY	Exhibition van.
NZ	Driving brake van (also known as driving van trailer).
QS	EMU translator vehicle.
YR	Ferry van (special Southern Region version of NJ with two pairs of side doors instead of three).

(2) A third letter denoting the brake type:

A	Air braked
V	Vacuum braked
X	Dual braked

OPERATING CODES

The normal operating codes are given in parentheses after the TOPS type codes. These are as follows:

BG	Gangwayed brake van.

BPOT	Brake post office stowage van.
DLV	Driving brake van (also known as driving van trailer – DVT).
GUV	General utility van.
PCV	Propelling control van.
POS	Post office sorting van.
POT	Post office stowage van.

AK51 (RK) KITCHEN CAR

Dia. AK503. Mark 1. Converted 1989 from RBR. Fluorescent lighting. Commonwealth bogies. ETH 2X.

Lot No. 30628 Pressed Steel 1960–61. 39 t.

Note: Kitchen cars have traditionally been numbered in the NPCCS series, but have passenger coach diagram numbers!

80041	(1690)	x	**CC**	RS	*ON*	BN

NN COURIER VEHICLE

Dia. NN504. Mark 1. Converted 1986–7 from BSKs. One compartment retained for courier use. Roller shutter doors. ETH 2.

80207. Lot No. 30721 Wolverton 1963. Commonwealth bogies. 37 t.
80211–7/23. Lot No. 30699 Wolverton 1962. Commonwealth bogies. 37 t.
80220. Lot No. 30573 Gloucester 1960. B4 bogies. 33 t.

Non-Standard Livery: 80211 is purple.

80207	(35466)	x	**PC**	VS	*ON*	SL
80211	(35296)		**0**	CN		FK
80212	(35307)	x	**RM**	E		OM
80213	(35316)	x	**CH**	PE		CP
80216	(35295)	x	**RM**	E		OM
80217	(35299)	x	**M**	14	*OS*	NY
80220	(35276)	x	**G**	WT		BQ
80223	(35331)	x	**RY**	CN		DY

Name: 80207 is branded 'BAGGAGE CAR No.11'.

NP POST OFFICE GUV

Dia. NP502. Mark 1. Converted 1991–93 from newspaper vans. Short frames (57'). Originally converted from GUV. Fluorescent lighting, toilet and gangways fitted. Load 14 t. B5 bogies. ETH 3X.

Lot No. 30922 Wolverton or Doncaster 1977–8. 31 t.

80251	(86467, 94017)	x	**RM**	E	TE
80252	(86718, 94022)		**RM**	E	TE
80253	(86170, 94018)		**RM**	E	OM
80254	(86082, 94012)	x	**RM**	E	OM
80255	(86098, 94019)	x	**RM**	E	OM
80256	(86408, 94013)	x	**RM**	E	OM
80257	(86221, 94023)	x	**RM**	E	OM

80258	(86651, 94002)		**RM**	E	OM
80259	(86845, 94005)	x	**RM**	E	OM

NS (POS) POST OFFICE SORTING VAN

Used in travelling post office (TPO) trains. Mark 1. Various diagrams.

The following lots have BR Mark 1 bogies except * B5 bogies. (subtract 2 t from weight).

80303–80305. Lot No. 30486 Wolverton 1959. Dia. NS501. Originally built with nets for collecting mail bags in motion. Equipment now removed. ETH 3X. 36 t.
80306–80308. Lot No. 30487 Wolverton 1959. Dia. NS502. ETH 3. 36 t.
80309–80314. Lot No. 30661 Wolverton 1961. Dia. NS501. ETH 3. 37 t.
80315–80316. Lot No. 30662 Wolverton 1961. Dia. NS501. ETH 3X. 36 t.

80303	x*	**RM**	E	OM
80305	v	**RM**	E	OM
80306	v	**RM**	E	OM
80308	x*	**RM**	E	OM
80309	x*	**RM**	E	OM
80310	v	**RM**	E	OM
80312	v	**RM**	E	OM
80314	x*	**RM**	E	OM
80315	v	**RM**	E	OM
80316	x*	**RM**	E	OM

The following lots a1re pressure ventilated and have B5 bogies.
80319–80327. Dia. NS504. Lot No. 30778 York 1968–9. ETH 4. 35 t.
80328–80338. Dia. NS505. Lot No. 30779 York 1968–9. ETH 4. 35 t.
80339–80355. Dia. NS506. Lot No. 30780 York 1968–9. ETH 4. 35 t.

80319		**RM**	E	*E*	EN
80320		**RM**	E	*E*	EN
80321		**RM**	E	*E*	BK
80322		**RM**	E	*E*	EN
80323		**RM**	E	*E*	EN
80324		**RM**	E	*E*	EN
80325		**RM**	E	*E*	EN
80326		**RM**	E	*E*	EN
80327		**RM**	E	*E*	BK
80328		**RM**	E		OM
80329		**RM**	E		OM
80330		**RM**	E		Pylle Hill (Bristol)
80331		**RM**	E	*E*	EN
80332		**RM**	E	*E*	EN
80333		**RM**	E	*E*	EN
80334		**RM**	E	*E*	BK
80335	x	**RM**	E		OM
80336		**RM**	E		Bristol E. Yd.
80337		**RM**	E	*E*	NC
80338		**RM**	E		Bristol E. Yd.
80339		**RM**	E	*E*	BK
80340		**RM**	E	*E*	BK
80341		**RM**	E	*E*	EN
80342		**RM**	E	*E*	BK
80343		**RM**	E	*E*	BK
80344		**RM**	E	*E*	BK
80345		**RM**	E	*E*	EN
80346		**RM**	E	*E*	EN
80347		**RM**	E	*E*	EN
80348		**RM**	E	*E*	BZ
80349		**RM**	E	*E*	EN
80350		**RM**	E	*E*	BK
80351		**RM**	E	*E*	EN
80352		**RM**	E	*E*	BK
80353		**RM**	E	*E*	EN
80354		**RM**	E	*E*	BK
80355		**RM**	E	*E*	EN

Names:

80320	The Borders Mail
80327	George James
80337	Brian Quinn

80356–80380. Lot No. 30839 York 1972–3. Dia. NS501. Pressure ventilated. Fluorescent lighting. B5 bogies. ETH 4X. 37 t.

80356	**RM**	E	*E*	BK
80357	**RM**	E	*E*	BK

80358 **RM** E *E* EN
80359 **RM** E *E* BK
80360 **RM** E *E* EN
80361 **RM** E *E* EN
80362 **RM** E *E* EN
80363 **RM** E *E* BK
80364 **RM** E *E* BK
80365 **RM** E *E* EN
80366 **RM** E *E* EN
80367 **RM** E *E* EN
80368 **RM** E *E* BK
80369 **RM** E *E* EN
80370 **RM** E *E* BK
80371 **RM** E *E* BZ
80372 **RM** E *E* EN
80373 **RM** E *E* EN
80374 **RM** E *E* EN
80375 **RM** E *E* BK
80376 **RM** E *E* BK
80377 **RM** E *E* EN
80378 **RM** E *E* EN
80379 **RM** E *E* EN
80380 **RM** E *E* BZ

Names:

80360 Derek Carter
80367 M.G. Berry

80381–80395. Lot No. 30900 Wolverton 1977. Dia NS531. Converted from SK. Pressure ventilated. Fluorescent lighting. B5 bogies. ETH 4X. 38 t.

80381 (25112) **RM** E *E* EN
80382 (25109) **RM** E *E* EN
80383 (25033) **RM** E *E* EN
80384 (25078) **RM** E *E* EN
80385 (25083) **RM** E *E* EN
80386 (25099) **RM** E *E* EN
80387 (25045) **RM** E *E* EN
80389 (25103) **RM** E ZG
80390 (25047) **RM** E *E* EN
80392 (25082) **RM** E *E* EN
80393 (25118) **RM** E *E* EN
80394 (25156) **RM** E *E* EN
80395 (25056) **RM** E *E* EN

Name:

80390 Ernie Gosling

NT (POT) POST OFFICE STOWAGE VAN

Mark 1. Open vans used for stowage of mail bags in conjunction with POS.

Lot No. 30488 Wolverton 1959. Dia. NT502. Originally built with nets for collecting mail bags in motion. Equipment now removed. B5 bogies. ETH 3. 35 t.

80400 **RM** E *E* BK
80401 **RM** E *E* EN
80402 **RM** E *E* BK

The following eight vehicles were converted at York from BSK to lot 30143 (80403) and 30229 (80404–80414). No new lot number was issued. Dia. NT503. B5 bogies. 35 t. (* Dia. NT501 BR2 bogies 38 t. ETH 3 (3X*).

80403 (34361) **RM** E *E* BZ
80404 (35014) **RM** E *E* BZ
80405 (35009) **RM** E *E* BZ
80406 (35022) **RM** E *E* EN
80411 (35003) * **RM** E *E* BZ
80412 (35002) * **RM** E *E* EN
80413 (35004) * **RM** E *E* EN
80414 (35005) * **RM** E *E* EN

Lot No. 30781 York 1968. Dia. NT505. Pressure ventilated. B5 bogies. ETH 4. 34 t.

80415 **RM** E *E* EN
80416 **RM** E *E* EN
80417 **RM** E *E* EN
80419 **RM** E *E* EN
80420 **RM** E *E* BK
80421 **RM** E *E* EN

80422	**RM**	E	*E*	EN	80424	**RM**	E	*E*	EN
80423	**RM**	E	*E*	EN					

Lot No. 30840 York 1973. Dia. NT504. Pressure ventilated. fluorescent lighting. B5 bogies. ETH 4X. 35 t.

80425	**RM**	E	*E*	BK	80428	**RM**	E	*E*	EN
80426	**RM**	E	*E*	EN	80429	**RM**	E	*E*	BK
80427	**RM**	E	*E*	EN	80430	**RM**	E	*E*	EN

Lot No. 30901 Wolverton 1977. converted from SK. Dia. NT521. Pressure ventilated. Fluorescent lighting. B5 bogies. ETH 4X. 35 t.

80431	(25104)	**RM**	E	*E*	EN	80436	(25077)	**RM**	E	*E*	EN
80432	(25071)	**RM**	E	*E*	EN	80437	(25068)	**RM**	E	*E*	EN
80433	(25150)	**RM**	E	*E*	BK	80438	(25139)	**RM**	E	*E*	BK
80434	(25119)	**RM**	E	*E*	EN	80439	(25127)	**RM**	E	*E*	BK
80435	(25117)	**RM**	E	*E*	EN						

NU (BPOT) BRAKE POST OFFICE STOWAGE VAN

Dia. NU502. Mark 1. As NT but with brake compartment. Pressure ventilated. B5 bogies. ETH4.

Lot No. 30782 York 1968. 36 t.

80456	**RM**	E	*E*	EN	80458	**RM**	E	*E*	EN
80457	**RM**	E	*E*	EN					

NZ (DLV) DRIVING BRAKE VAN (110 m.p.h.)

Dia. NZ501. Mark 3B. Air conditioned. T4 bogies. dg. ETH 5X.

Lot No. 31042 Derby 1988. 45.18 t.

82101	**V**	P	*VW*	OY	82121	**V**	P	*VW*	PC
82102		P	*VW*	OY	82122	**V**	P	*VW*	MA
82103	**V**	P	*VW*	OY	82123	**V**	P	*VW*	PC
82104	**V**	P	*VW*	PC	82124		P	*VW*	PC
82105	**V**	P	*VW*	PC	82125	**V**	P	*VW*	PC
82106	**V**	P	*VW*	OY	82126	**V**	P	*VW*	OY
82107		P	*VW*	PC	82127	**V**	P	*VW*	OY
82108	**V**	P	*VW*	PC	82128	**V**	P	*VW*	OY
82109	**V**	P	*VW*	PC	82129	**V**	P	*VW*	OY
82110	**V**	P	*VW*	PC	82130	**V**	P	*VW*	MA
82111	**V**	P	*VW*	PC	82131	**V**	P	*VW*	OY
82112	**V**	P	*VW*	PC	82132	**V**	P	*VW*	OY
82113	**V**	P	*VW*	OY	82133	**V**	P	*VW*	OY
82114		P	*VW*	PC	82134	**V**	P	*VW*	OY
82115	**V**	P	*VW*	MA	82135	**V**	P	*VW*	MA
82116		P	*VW*	PC	82136		P	*VW*	MA
82117	**V**	P	*VW*	PC	82137	**V**	P	*VW*	MA
82118		P	*VW*	OY	82138	**V**	P	*VW*	PC
82119	**V**	P	*VW*	MA	82139	**V**	P	*VW*	PC
82120	**V**	P	*VW*	MA	82140	**V**	P	*VW*	MA

82141	**V**	P	*VW*	MA	82147	**V**	P	*VW*	MA
82142	**V**	P	*VW*	MA	82148	**V**	P	*VW*	PC
82143		P	*VW*	OY	82149	**V**	P	*VW*	PC
82144	**V**	P	*VW*	OY	82150	**V**	P	*VW*	PC
82145	**V**	P	*VW*	MA	82151		P	*VW*	OY
82146	**V**	P	*VW*	MA	82152	**V**	P	*VW*	MA

Names:

82115 Liverpool John Moores University
82120 Liverpool Chamber of Commerce
82121 Carlisle Cathedral
82124 The Girls' Brigade
82126 G8 Summit Birmingham 1998
82127 Abraham Darby
82132 INDUSTRY 96 West Midlands
82134 Sir Henry Doulton 1820–1897
82135 Spirit of Cumbria
82147 The Red Devils
82148 International Spring Fair
82149 101 Squadron

NZ (DLV) DRIVING BRAKE VAN (140 m.p.h.)

Dia. NZ502. Mark 4. Air conditioned. Swiss-built (SIG) bogies. dg. ETH 6X.

Lot No. 31043 Metro-Cammell 1988. 45.18 t.

82200	**GN**	H	*GN*	BN	82216	**GN**	H	*GN*	BN
82201	**GN**	H	*GN*	BN	82217	**GN**	H	*GN*	BN
82202	**GN**	H	*GN*	BN	82218	**GN**	H	*GN*	BN
82203	**GN**	H	*GN*	BN	82219	**GN**	H	*GN*	BN
82204	**GN**	H	*GN*	BN	82220	**GN**	H	*GN*	BN
82205	**GN**	H	*GN*	BN	82221	**GN**	H	*GN*	BN
82206	**GN**	H	*GN*	BN	82222	**GN**	H	*GN*	BN
82207	**GN**	H	*GN*	BN	82223	**GN**	H	*GN*	BN
82208	**GN**	H	*GN*	BN	82224	**GN**	H	*GN*	BN
82209	**GN**	H	*GN*	BN	82225	**GN**	H	*GN*	BN
82210	**GN**	H	*GN*	BN	82226	**GN**	H	*GN*	BN
82211	**GN**	H	*GN*	BN	82227	**GN**	H	*GN*	BN
82212	**GN**	H	*GN*	BN	82228	**GN**	H	*GN*	BN
82213	**GN**	H	*GN*	BN	82229	**GN**	H	*GN*	BN
82214	**GN**	H	*GN*	BN	82230	**GN**	H	*GN*	BN
82215	**GN**	H	*GN*	BN	82231	**GN**	H	*GN*	BN

ND (BG) GANGWAYED BRAKE VAN (90 m.p.h.)

Dia. ND501. Mark 1. Short frames (57'). Load 10t. All vehicles were built with BR Mark 1 bogies. ETH 1. Vehicles numbered 81xxx had 3000 added to the original numbers to avoid confusion with Class 81 locomotives. The full lot number list is listed here for reference purposes with renumbered vehicles. No unmodified vehicles remain in service.

80525. Lot No. 30009 Derby 1952–3. 31 t.
80621. Lot No. 30046 York 1954. 31.5 t.
80700. Lot No. 30136 Metro-Cammell 1955. 31.5 t.
80731–80791. Lot No. 30140 BRCW 1955–6. 31.5 t.
80826–80848. Lot No. 30144 Cravens 1955. 31.5 t.
80855–80960. Lot No. 30162 Pressed Steel 1956–7. 32 t.
80971–81014. Lot No. 30173 York 1956. 31.5 t.
81025–81051. Lot No. 30224 Cravens 1956. 31.5 t.
81055–81175. Lot No. 30228 Metro-Cammell 1957–8. 31.5 t.
81182–81188. Lot No. 30234 Cravens 1956–7. 31.5 t.
81205–81265. Lot No. 30163 Pressed Steel 1957. 31.5 t.
81266–81309. Lot No. 30323 Pressed Steel 1957. 32 t.
81316–81497. Lot No. 30400 Pressed Steel 1957–8. 32 t.
81498–81568. Lot No. 30484 Pressed Steel 1958. 32 t.
81590. Lot No. 30715 Gloucester 1962. 31 t.
81604–81606. Lot No. 30716 Gloucester 1962. 31 t.

The following converted NDs are in service:

84025. ND rebogied with Commonwealth bogies (add 1.5 t to weight) and adapted for use as exhibition van 1998 at Lancastrian Carriage & Wagon Co. Ltd.
84382/7/477. Dia. NB501. High security brake van. Converted 1985 at Wembley Heavy Repair Depot from ND. Gangways removed. B4 bogies.

84025	(81025)	v	**M**	RA	CP
84382	(81382, 80460)	x	**RX**	E	Cambridge Coalfields Sdgs.
84387	(81387, 80461)	x	**B**	E	Crewe South Yard
84477	(81477, 80463)	x	**B**	E	Crewe South Yard

NJ (GUV) GENERAL UTILITY VAN

Dia. NJ501. Mark 1. Short frames. Load 14 t. Screw couplings. These vehicles had 7000 added to the original numbers to avoid confusion with Class 86 locomotives. The full lot number list is listed here for reference purposes with renumbered vehicles. No unmodified vehicles remain in service. All vehicles were built with BR Mark 2 bogies. ETH 0 or 0X*.

86081–86499. Lot No. 30417 Pressed Steel 1958–9. 30 t.
86508–86518. Lot No. 30343 York 1957. 30 t.
86521–86651. Lot No. 30403 Glasgow 1958–60. 30 t.
86656–86834. Lot No. 30565 Pressed Steel 1959. 30 t.
86836–86978. Lot No. 30616 Pressed Steel 1959–60. 30 t.

NE/NH (BG) GANGWAYED BRAKE VAN (100/110 m.p.h.)

NE are ND but rebogied with B4 bogies suitable for 100 m.p.h. NH are identical but are allowed to run at 110 m.p.h. with special maintenance of the bogies. For lot numbers refer to original number series. Deduct 1.5t from weights. All NHA are *pg. ETH 1 (1X*).

Non-standard Livery: 92116 is purple.

92100	(81391)	to	RV	CP
92111	(81432)	NHA	H	CP

92114	(81443)	NHA		H		LT
92116	(81450)	to	**0**	CN		FK
92125	(81470)	to		DR		SD
92146	(81498)	NHA		H		LT
92159	(81534)	NHA		H	*SR*	IS
92174	(81567)	NHA		H	*SR*	IS
92175	(81568)	pg		H		CP
92193	(81604)	pg		E		Preston Station
92194	(81606)	to		H	*SR*	IS
92211	(81267)	*	**R**	E		BK
92229	(80902)	*	**R**	E		BK
92234	(81336, 84336)	*	**RX**	E		DY
92238	(81563, 84563)		**RY**	E		DY
92252	(80959)	x*	**RY**	E		BK
92261	(80988)	x*	**RY**	E		BK
92267	(81404, 84404)	x		E		BK

NE (BG) GANGWAYED BRAKE VAN (100 m.p.h.)

As ND but rebogied with Commonwealth bogies suitable for 100 m.p.h. ETH 1 (1X*). For lot numbers refer to original number series. Add 1.5 t to weights to allow for the increased weight of the Commonwealth bogies.

92302	(81501, 84501)		**RX**	E	KM
92303	(81427, 84427)		**RX**	E	DY
92306	(81217, 84217)	*	**RY**	E	KM
92309	(81043, 84043)	x*	**RX**	E	KM
92311	(81453, 84453)	x	**RY**	E	Crewe South Yard
92312	(81548, 84548)		**RX**	E	KM
92314	(80777)	x*	**RY**	E	Crewe South Yard
92316	(80980)	x*	**RY**	E	KM
92319	(81055, 84055)	*	**RY**	E	KM
92321	(81566, 84566)		**RY**	E	FK
92323	(80832)	*	**R**	E	KM
92324	(81087, 84087)		**RY**	E	KM
92325	(80791)		**RY**	E	KM
92328	(80999)	x*	**RY**	E	KM
92329	(81001, 84001)	*	**RY**	E	KM
92330	(80995)	x*	**RY**	E	KM
92332	(80845)	*	**RX**	E	KM
92333	(80982)	*	**RY**	E	KM
92337	(81140, 84140)	*	**RX**	E	KM
92340	(81059, 84059)	*	**RY**	E	KM
92341	(81316, 84316)	x	**RY**	E	KM
92343	(81505, 84505)	x	**R**	E	KM
92344	(81154, 84154)	*	**RY**	E	KM
92345	(81083, 84083)	x*	**RY**	E	KM
92346	(81091, 84091)		**RY**	E	KM
92347	(81326, 84326)		**RX**	E	DY
92348	(81075, 84075)	x*	**R**	E	KM
92350	(81049, 84049)	*	**RY**	E	DY

92353	(81323, 84323)		**R**	E		KM
92355	(81517, 84517)	x	**RX**	E		DY
92356	(81535, 84535)	x		E		KM
92357	(81136, 84136)		**RX**	E		KM
92362	(81188, 84188)	x	**RY**	E		KM
92363	(81294, 84294)	x	**RY**	E		Crewe South Yard
92364	(81030, 84030)	x*	**R**	E		KM
92365	(81122, 84122)		**RX**	E		KM
92366	(81551, 84551)		**RX**	E		KM
92369	(80960)	x*		E		Doncaster Dock Siding
92370	(81324, 84324)		**RX**	E		KM
92377	(80928)	*	**RX**	E		DY
92379	(80914)	*	**RX**	E		KM
92380	(81247, 84247)	*	**R**	E		KM
92381	(81476, 84476)		**RX**	E		KM
92382	(81561, 84561)		**RX**	E		DY
92384	(80893)		**RY**	E		Crewe South Yard
92385	(81261, 84261)	x*	**RY**	E		KM
92389	(81026, 84026)	*	**RY**	E		KM
92390	(80834)	*		E		KM
92392	(80861)	*	**RY**	E		KM
92395	(81274, 84274)			E		KM
92398	(80859)	x*	**RY**	E		KM
92400	(81211, 84211)	*		E		Crewe South Yard
92401	(81280, 84280)	x	**RX**	E		KM
92402	(81099, 84099)	*	**RY**	E		KM
92403	(81273, 84273)	x	**RY**	E		OM
92404	(81051, 84051)	x		E		KM
92409	(81370, 84370)	x		E		OM
92410	(81469, 84469)	x		E		Crewe South Yard
92412	(81354, 84354)	*	**RY**	E		Crewe South Yard
92413	(81472, 84472)	x	**RY**	E		Crewe South Yard
92414	(81458, 84458)	x		E		OM
92415	(81388, 84388)		**RX**	E		KM
92416	(81250, 84250)	*	**RY**	E		KM
92417	(80885)	*	**RX**	E		KM
92418	(81512, 84512)	*	**RX**	E		OM

NF (BG) GANGWAYED BRAKE VAN (100 m.p.h.)

As NE but with emergency equipment removed. For details and lot numbers refer to original number series. 92518–92728 have B4 bogies whilst 92804–92897 have Commonwealth bogies.

b (Dia. NB501). High security brake van. Converted at Wembley Heavy Repair Depot from ND 1985. Gangways removed. Now used for movement of materials between EWS maintenance depots.

92518	(80941, 92918)		**RX**	E		BK
92530	(81461, 84461)	xb	**RX**	E	*E*	EN
92542	(81207, 92942)		**RX**	E		KM
92562	(81232, 92962)		**RX**	E		Willesden Brent Sidings

92568	(81244, 92968)		**RX**	E		OM
92607	(81410, 92107)		**RX**	E		OM
92649	(81509, 92149)	x		E		BK
92728	(80921, 92228)		**RX**	E		BK
92804	(81339, 92304)	x	**RX**	E		KM
92805	(81590, 92305)	x	**RX**	E		KM
92810	(81105, 92310)		**RX**	E		KM
92815	(80848, 92315)	*	**RX**	E		CU
92817	(80836, 92317)	x	**RX**	E		KM
92822	(80771, 92322)	x	**RX**	E		KM
92827	(80842, 92327)	x	**RX**	E		KM
92831	(81365, 92331)	x	**RX**	E		KM
92842	(81397, 92342)	x	**RY**	E		KM
92852	(81182, 92352)	*	**RX**	E		DY
92854	(81353, 92354)	x	**RX**	E		KM
92859	(81275, 92359)	*	**RX**	E		DY
92860	(81431, 92360)		**RX**	E		Willesden Brent Sdgs.
92861	(81463, 92361)		**R**	E		KM
92867	(81293, 92367)	x	**RX**	E		CU
92872	(81362, 92372)	x	**RY**	E		Crewe South Yard
92873	(81528, 92373)		**RX**	E		KM
92876	(81374, 92376)	*	**RX**	E		KM
92883	(81429, 92383)	*	**RX**	E		OM
92886	(80843, 92386)	x	**RX**	E		KM
92897	(80700, 92397)	x*	**RY**	E		KM

NE/NH (BG) GANGWAYED BRAKE VAN (100/110 m.p.h.)

Renumbered from 920xx series by adding 900 to number to avoid conflict with Class 92 locos. Class continued from 92267.

92901	(80855, 92001)	NHA		H	*SR*	IS
92904	(80867, 92004)	*pg	**G**	VS		SL
92908	(80895, 92008)	NHA		H	*SR*	IS
92912	(80910, 92012)	*pg		H		KN
92916	(80930, 92016)	x*pg	**RY**	E		BK
92923	(80971, 92023)	*pg		H		LT
92927	(81061, 92027)	NHA		H		LT
92928	(81064, 92028)	NHA		H		LT
92929	(81077, 92029)	NHA		H	*GW*	OO
92931	(81102, 92031)	NHA		H	*SR*	IS
92933	(81123, 92033)	NHA		H		ZC
92934	(81142, 92034)	NHA		H		LT
92935	(81150, 92035)	*pg		H		ZC
92936	(81158, 92036)	NHA		H	*SR*	IS
92937	(81165, 92037)	NHA		H		ZH
92938	(81173, 92038)	NHA		H	*SR*	IS
92939	(81175, 92039)	NHA		H		ZH
92940	(81186, 92040)	pg		H	*SR*	IS
92946	(81214, 92046)	NHA		H	*SR*	IS
92948	(81218, 92048)	NHA		H		ZC

92961	(81231, 92061)		H	LT
92986	(81282, 92086)	to	H	CP
92988	(81284, 92088)	to	H	LT
92991	(81308, 92091)	to	H	LT
92998	(81381, 92098)	NHA	H	LT

NL NEWSPAPER VAN

Dia. NL501. Mark 1. Short frames (57'). Converted from NJ (GUV). Fluorescent lighting, toilets and gangways fitted. Load 14 t. As EWS does not now carry newspaper traffic these are now all out of use. B5 Bogies. ETH 3X.

Lot No. 30922 Wolverton or Doncaster 1977–8. 31 t.

94003	(86281, 93999)	x	**RX**	E	OM
94004	(86156, 85504)		**RY**	E	OM
94006	(86202, 85506)		**RX**	E	OM
94007	(86572, 85507)		**B**	E	OM
94009	(86144, 85509)		**RY**	E	OM
94010	(86151, 85510)	x	**RX**	E	OM
94011	(86437, 85511)		**RX**	E	OM
94016	(86317, 85516)	x	**B**	E	OM
94020	(86220, 85520)	x	**RY**	E	OM
94021	(86204, 85521)	x	**B**	E	OM
94024	(86106, 85524)		**B**	E	OM
94025	(86377, 85525)		**RY**	E	TE
94026	(86703, 85526)	x	**RY**	E	OM
94027	(86732, 85527)		**R**	E	FK
94028	(86733, 85528)	x	**RX**	E	TE
94029	(86740, 85529)	x	**RY**	E	OM
94030	(86746, 85530)	x	**B**	E	OM
94032	(86730, 85532)		**RX**	E	OM

NKA HIGH SECURITY GENERAL UTILITY VAN

Dia. NK501. Mark 1. These vehicles are GUVs further modified with new floors, three roller shutter doors per side and the end doors removed. For lot Nos. see original number series. Commonwealth bogies. Add 2 t to weight. ETH0X.

94100	(86668, 95100)	**RX**	E	*E*	BK
94101	(86142, 95101)	**RX**	E	*E*	BK
94102	(86762, 95102)	**RX**	E	*E*	BK
94103	(86956, 95103)	**RX**	E	*E*	BK
94104	(86942, 95104)	**RX**	E	*E*	EN
94106	(86353, 95106)	**RX**	E	*E*	BK
94107	(86576, 95107)	**RX**	E	*E*	EN
94108	(86600, 95108)	**RX**	E	*E*	BK
94110	(86393, 95110)	**RX**	E	*E*	BK
94111	(86578, 95111)	**RX**	E	*E*	EN
94112	(86673, 95112)	**RX**	E	*E*	EN
94113	(86235, 95113)	**RX**	E	*E*	BK
94114	(86081, 95114)	**RX**	E	*E*	BK

94116	(86426, 95116)	**RX**	E	*E*	BK
94117	(86534, 95117)	**RX**	E	*E*	BK
94118	(86675, 95118)	**RX**	E	*E*	EN
94119	(86167, 95119)	**RX**	E	*E*	EN
94121	(86518, 95121)	**RX**	E	*E*	BK
94123	(86376, 95123)	**RX**	E	*E*	BK
94126	(86692, 95126)	**RX**	E	*E*	BK
94132	(86607, 95132)	**RX**	E	*E*	EN
94133	(86604, 95133)	**RX**	E	*E*	BK
94137	(86610, 95137)	**RX**	E	*E*	EN
94138	(86212, 95138)	**RX**	E	*E*	EN
94140	(86571, 95140)	**RX**	E	*E*	BK
94146	(86648, 95146)	**RX**	E	*E*	BK
94147	(86091, 95147)	**RX**	E	*E*	BK
94148	(86416, 95148)	**RX**	E	*E*	EN
94150	(86560, 95150)	**RX**	E	*E*	BK
94153	(86798, 95153)	**RX**	E	*E*	EN
94155	(86820, 95155)	**RX**	E	*E*	EN
94157	(86523, 95157)	**RX**	E	*E*	EN
94160	(86581, 95160)	**RX**	E	*E*	BK
94164	(86104, 95164)	**RX**	E	*E*	EN
94166	(86112, 95166)	**RX**	E	*E*	BK
94168	(86914, 95168)	**RX**	E	*E*	BK
94170	(86395, 95170)	**RX**	E	*E*	BK
94172	(86429, 95172)	**RX**	E	*E*	EN
94174	(86852, 95174)	**RX**	E	*E*	EN
94175	(86521, 95175)	**RX**	E	*E*	BK
94176	(86210, 95176)	**RX**	E	*E*	EN
94177	(86411, 95177)	**RX**	E	*E*	BK
94180	(86362, 95141)	**RX**	E	*E*	EN
94182	(86710, 95182)	**RX**	E	*E*	BK
94190	(86624, 95350)	**RX**	E	*E*	EN
94191	(86596, 95351)	**RX**	E	*E*	BK
94192	(86727, 95352)	**RX**	E	*E*	EN
94193	(86514, 95353)	**RX**	E	*E*	EN
94195	(86375, 95355)	**RX**	E	*E*	BK
94196	(86478, 95356)	**RX**	E	*E*	BK
94197	(86508, 95357)	**RX**	E	*E*	BK
94198	(86195, 95358)	**RX**	E	*E*	BK
94199	(86854, 95359)	**RX**	E	*E*	BK
94200	(86207, 95360)	**RX**	E	*E*	EN
94202	(86563, 95362)	**RX**	E	*E*	BK
94203	(86345, 95363)	**RX**	E	*E*	BK
94204	(86715, 95364)	**RX**	E	*E*	BK
94205	(86857, 95365)	**RX**	E	*E*	BK
94207	(86529, 95367)	**RX**	E	*E*	BK
94208	(86656, 95368)	**RX**	E	*E*	EN
94209	(86390, 95369)	**RX**	E	*E*	BK
94211	(86713, 95371)	**RX**	E	*E*	EN
94212	(86728, 95372)	**RX**	E	*E*	EN
94213	(86258, 95373)	**RX**	E	*E*	EN

94214	(86367, 95374)	**RX**	E	*E*	BK
94215	(86862, 94077)	**RX**	E	*E*	BK
94216	(86711, 93711)	**RX**	E	*E*	EN
94217	(86131, 93131)	**RX**	E	*E*	BK
94218	(86541, 93541)	**RX**	E	*E*	BK
94221	(86905, 93905)	**RX**	E	*E*	BK
94222	(86474, 93474)	**RX**	E	*E*	EN
94223	(86660, 93660)	**RX**	E	*E*	BK
94224	(86273, 93273)	**RX**	E	*E*	BK
94225	(86849, 93849)	**RX**	E	*E*	BK
94226	(86525, 93525)	**RX**	E	*E*	BK
94227	(86585, 93585)	**RX**	E	*E*	BK
94228	(86511, 93511)	**RX**	E	*E*	BK
94229	(86720, 93720)	**RX**	E	*E*	BK

NAA PROPELLING CONTROL VEHICLE

Dia. NA508. Mark 1. Class 307 driving trailers converted for use in propelling mail trains out of termini. Fitted with roller shutter doors. Equipment fitted for communication between cab of PCV and locomotive. B5 bogies. ETH 2X.

Lot No. 30206 Eastleigh 1954–6. Converted at Hunslet-Barclay, Kilmarnock 1994–6.

94302	(75124)	**RX**	E	*E*	BK
94303	(75131)	**RX**	E	*E*	EN
94304	(75107)	**RX**	E	*E*	EN
94305	(75104)	**RX**	E	*E*	EN
94306	(75112)	**RX**	E	*E*	BK
94307	(75127)	**RX**	E	*E*	EN
94308	(75125)	**RX**	E	*E*	EN
94309	(75130)	**RX**	E	*E*	EN
94310	(75119)	**RX**	E	*E*	EN
94311	(75105)	**RX**	E	*E*	EN
94312	(75126)	**RX**	E	*E*	BK
94313	(75129)	**RX**	E	*E*	BK
94314	(75109)	**RX**	E	*E*	BK
94315	(75132)	**RX**	E	*E*	EN
94316	(75108)	**RX**	E	*E*	EN
94317	(75117)	**RX**	E	*E*	EN
94318	(75115)	**RX**	E	*E*	EN
94319	(75128)	**RX**	E	*E*	EN
94320	(75120)	**RX**	E	*E*	EN
94321	(75122)	**RX**	E	*E*	EN
94322	(75111)	**RX**	E	*E*	BK
94323	(75110)	**RX**	E	*E*	EN
94324	(75103)	**RX**	E	*E*	EN
94325	(75113)	**RX**	E	*E*	EN
94326	(75123)	**RX**	E	*E*	BK
94327	(75116)	**RX**	E	*E*	EN
94331	(75022)	**RX**	E	*E*	BK
94332	(75011)	**RX**	E	*E*	EN
94333	(75016)	**RX**	E	*E*	EN
94334	(75017)	**RX**	E	*E*	EN
94335	(75032)	**RX**	E	*E*	BK
94336	(75031)	**RX**	E	*E*	EN
94337	(75029)	**RX**	E	*E*	EN
94338	(75008)	**RX**	E	*E*	EN
94339	(75024)	**RX**	E	*E*	BK
94340	(75012)	**RX**	E	*E*	BK
94341	(75007)	**RX**	E	*E*	BK
94342	(75005)	**RX**	E	*E*	BK
94343	(75027)	**RX**	E	*E*	EN
94344	(75014)	**RX**	E	*E*	BK
94345	(75004)	**RX**	E	*E*	EN

NBA HIGH SECURITY BRAKE VAN (100 m.p.h.)

Dia. NB501. Mark 1. These vehicles are NEs further modified with sealed gangways, new floors, built-in tail lights and roller shutter doors. For lot Nos. see original number series. B4 bogies. 31.4 t. ETH 1X.

94400	(81224, 92954)	**RX**	E	*E*	BK
94401	(81277, 92224)	**RX**	E	*E*	EN
94403	(81479, 92629)	**RX**	E	*E*	BK
94404	(81486, 92135)	**RX**	E	*E*	BK
94405	(80890, 92233)	**RX**	E	*E*	EN
94406	(81226, 92956)	**RX**	E	*E*	BK
94407	(81223, 92553)	**RX**	E	*E*	BK
94408	(81264, 92981)	**RX**	E	*E*	BK
94410	(81205, 92941)	**RX**	E	*E*	EN
94411	(81378, 92997)	**RX**	E	*E*	EN
94412	(81210, 92945)	**RX**	E	*E*	EN
94413	(80909, 92236)	**RX**	E	*E*	NC
94414	(81377, 92996)	**RX**	E	*E*	BK
94415	(81309, 92992)	**RX**	E	*E*	EN
94416	(80929, 92746)	**RX**	E	*E*	BK
94418	(81248, 92244)	**RX**	E	*E*	BK
94419	(80858, 92902)	**RX**	E	*E*	BK
94420	(81325, 92263)	**RX**	E	*E*	BK
94421	(81230, 92960)	**RX**	E	*E*	EN
94422	(81516, 92651)	**RX**	E	*E*	BK
94423	(80923, 92914)	**RX**	E	*E*	BK
94424	(81400, 92103)	**RX**	E	*E*	EN
94425	(80937, 92212)	**RX**	E	*E*	BK
94426	(81283, 92987)	**RX**	E	*E*	BK
94427	(80894, 92754)	**RX**	E	*E*	BK
94428	(81550, 92166)	**RX**	E	*E*	BK
94429	(80870, 92232)	**RX**	E	*E*	EN
94430	(80908, 92235)	**RX**	E	*E*	BZ
94431	(81401, 92604)	**RX**	E	*E*	BK
94432	(81383, 92999)	**RX**	E	*E*	EN
94433	(81495, 92643)	**RX**	E	*E*	EN
94434	(81268, 92584)	**RX**	E	*E*	NC
94435	(81485, 92134)	**RX**	E	*E*	EN
94436	(81237, 92565)	**RX**	E	*E*	BK
94437	(81403, 92208)	**RX**	E	*E*	EN
94438	(81425, 92251)	**RX**	E	*E*	BK
94439	(81480, 92130)	**RX**	E	*E*	BK
94440	(81497, 92645)	**RX**	E	*E*	BK
94441	(81492, 92140)	**RX**	E	*E*	BK
94442	(80932, 92723)	**RX**	E	*E*	BZ
94443	(81473, 92127)	**RX**	E	*E*	BK
94444	(81484, 92133)	**RX**	E	*E*	BK
94445	(81444, 92615)	**RX**	E	*E*	EN
94446	(80857, 92242)	**RX**	E	*E*	EN
94447	(81515, 92266)	**RX**	E	*E*	EN
94448	(81541, 92664)	**RX**	E	*E*	BK
94449	(81536, 92747)	**RX**	E	*E*	BK
94450	(80927, 92915)	**RX**	E	*E*	BK
94451	(80955, 92257)	**RX**	E	*E*	EN
94452	(81394, 92602)	**RX**	E	*E*	BK
94453	(81170, 92239)	**RX**	E	*E*	EN

94454	(81465, 92124)	**RX**	E	*E*	EN
94455	(81239, 92264)	**RX**	E	*E*	BK
94457	(81454, 92119)	**RX**	E	*E*	BK
94458	(81255, 92974)	**RX**	E	*E*	BK
94459	(81490, 92138)	**RX**	E	*E*	BK
94460	(81266, 92983)	**RX**	E	*E*	EN
94461	(81487, 92136)	**RX**	E	*E*	EN
94462	(81289, 92270)	**RX**	E	*E*	EN
94463	(81375, 92995)	**RX**	E	*E*	EN
94464	(81240, 92262)	**RX**	E	*E*	EN
94465	(81481, 92131)	**RX**	E	*E*	BK
94466	(81236, 92964)	**RX**	E	*E*	EN
94467	(81245, 92969)	**RX**	E	*E*	BK
94468	(81259, 92978)	**RX**	E	*E*	BK
94469	(81260, 92979)	**RX**	E	*E*	BK
94470	(81442, 92113)	**RX**	E	*E*	NC
94471	(81518, 92152)	**RX**	E	*E*	BK
94472	(81526, 92975)	**RX**	E	*E*	BK
94473	(81262, 92272)	**RX**	E	*E*	EN
94474	(81452, 92618)	**RX**	E	*E*	EN
94475	(81208, 92943)	**RX**	E	*E*	EN
94476	(81209, 92944)	**RX**	E	*E*	BK
94477	(81494, 92642)	**RX**	E	*E*	BK
94478	(81488, 92637)	**RX**	E	*E*	EN
94479	(81482, 92132)	**RX**	E	*E*	BK
94480	(81411, 92608)	**RX**	E	*E*	EN
94481	(81493, 92641)	**RX**	E	*E*	BK
94482	(81491, 92639)	**RX**	E	*E*	NC
94483	(81500, 92647)	**RX**	E	*E*	EN
94484	(81426, 92110)	**RX**	E	*E*	BK
94485	(81496, 92644)	**RX**	E	*E*	EN
94486	(81254, 92973)	**RX**	E	*E*	BK
94487	(81413, 92609)	**RX**	E	*E*	BK
94488	(81405, 92105)	**RX**	E	*E*	BK
94489	(81423, 92230)	**RX**	E	*E*	BK
94490	(81409, 92606)	**RX**	E	*E*	EN
94491	(80936, 92753)	**RX**	E	*E*	BK
94492	(80888, 92721)	**RX**	E	*E*	BK
94493	(80944, 92919)	**RX**	E	*E*	BK
94494	(81451, 92617)	**RX**	E	*E*	BK
94495	(80871, 92755)	**RX**	E	*E*	BK
94496	(81514, 92650)	**RX**	E	*E*	BK
94497	(80877, 92717)	**RX**	E	*E*	BK
94498	(81225, 92555)	**RX**	E	*E*	BK
94499	(81258, 92577)	**RX**	E	*E*	BK
94500	(81457, 92121)	**RX**	E	*E*	BK

NBA/NIA HIGH SECURITY BRAKE VAN (100/110 m.p.h.)

Dia. NB501 or NI501. Mark 1. These vehicles are NEs further modified with sealed gangways, new floors, built-in tail lights and roller shutter doors. For lot Nos. see original number series. B4 bogies. 31.4 t. ETH 1X.

These vehicles are identical to the 94400–94500 series. Certain vehicles are being given a special maintenance regime whereby tyres are reprofiled more frequently than normal and are then allowed to run at 110 m.p.h. Vehicles from the 94400 series upgraded to 110 m.p.h. are being renumbered in this series. Vehicles are NBA (100 m.p.h.) unless marked NIA (110 m.p.h.)

94501	(80891, 92725)		**RX**	E	*E*	BK
94502	(80924, 92720)	NIA	**RX**	E	*E*	BK
94503	(80873, 92709)	NIA	**RX**	E	*E*	ML
94504	(80935, 92748)		**RX**	E	*E*	BK
94505	(81235, 92750)	NIA	**RX**	E	*E*	ML
94506	(80958, 92922)	NIA	**RX**	E	*E*	ML
94507	(80876, 92505)	NIA	**RX**	E	*E*	BK
94508	(80887, 92722)	NIA	**RX**	E	*E*	BK
94509	(80897, 92509)		**RX**	E	*E*	BK
94510	(80945, 92265)		**RX**	E	*E*	BK
94511	(81504, 92714)	NIA	**RX**	E	*E*	ML
94512	(81265, 92582)		**RX**	E	*E*	BK
94513	(81257, 92576)		**RX**	E	*E*	BK
94514	(81459, 92122)	NIA	**RX**	E	*E*	ML
94515	(80916, 92513)	NIA	**RX**	E	*E*	ML
94516	(,)		**RX**	E		
94517	(81489, 92243)	NIA	**RX**	E	*E*	ML
94518	(81346, 92258)		**RX**	E	*E*	BK
94519	(,)		**RX**	E		
94520	(80940, 92917)	NIA	**RX**	E	*E*	ML
94521	(80900, 92510)	NIA	**RX**	E	*E*	ML
94522	(80880, 92907)	NIA	**RX**	E	*E*	ML
94523	(,)		**RX**	E		
94524	(,)		**RX**	E		
94525	(,)		**RX**	E		
94526	(,)		**RX**	E		
94527	(,)		**RX**	E		
94528	(,)		**RX**	E		
94529	(,)		**RX**	E		
94530	(81511, 94409)	NIA	**RX**	E	*E*	BK
94531	(80879, 94456)	NIA	**RX**	E	*E*	ML
94532	(,)		**RX**	E		
94533	(,)		**RX**	E		
94534	(,)		**RX**	E		
94535	(,)		**RX**	E		
94536	(,)		**RX**	E		
94537	(,)		**RX**	E		

NO (GUV) GENERAL UTILITY VAN (100 m.p.h.)

Dia. NO513. Mark 1. For lot Nos. see original number series. Commonwealth bogies except where shown otherwise. Add 2 t to weight (Subtract 1 t for B4). ETH 0X.

95105	(86126, 93126)		**RX**	E	CU
95109	(86269, 93269)	x	**B**	E	CU
95120	(86468, 93468)	x	**RY**	E	CU
95124	(86836, 93836)	x	**R**	E	CU
95125	(86143, 93143)	x	**B**	E	CU
95128	(86764, 93764)	x	**RY**	E	Crewe South Yard
95129	(86347, 93347)	x	**RY**	E	Crewe South Yard
95131	(86860, 93860)		**RX**	E	OM
95135	(86249, 93249)	x	**RY**	E	CU
95136	(86396, 93396)	x	**RX**	E	OM
95144	(86165, 93165)	x	**RY**	E	OM
95145	(86293, 93293)	x	**RX**	E	CU
95151	(86606, 93606)	x	**RX**	E	OM
95152	(86969, 93969)	x	**RY**	E	CU
95156	(86160, 93160)	x	**RX**	E	OM
95165	(86262, 93262)	x	**RX**	E	CU
95167	(86255, 93255)		**RX**	E	OM
95171	(86110, 93110)	x	**RX**	E	OM
95190	(86643, 95393)	B4	**RY**	E	OM
95191	(86278, 95391)	x B4	**B**	E	OM
95194	(86192, 93192)	x B4	**RX**	E	OM
95195	(86539, 93539)	x B4	**RX**	E	OM
95196	(86775, 93775)	x B4	**RX**	E	OM
95197	(86590, 93590)	x B4	**RX**	E	OM
95198	(86134, 93134)	x B4	**RX**	E	OM
95199	(86141, 93141)	x B4	**RX**	E	OM

NCX NEWSPAPER VAN (100 m.p.h.)

Dia. NC501. Mark 1. BGs modified to carry newspapers. As EWS does not now carry newspaper traffic these are now all out of use. For lot Nos. refer to original number series. Commonwealth bogies. Add 2 t to weight. ETH3 (3X*).

95201	(80875)	x	**RX**	E	KM
95204	(80947)	x*	**RX**	E	OM
95210	(80731)	x	**RX**	E	OM
95211	(80949)	x	**RY**	E	KM
95217	(81385, 84385)	x	**B**	E	OM
95223	(80933)	x*	**RY**	E	OM
95227	(81292, 95310)	x	**RX**	E	KM
95228	(81014, 95332)	x	**RX**	E	NC
95229	(81341, 95329)	x	**RX**	E	OM
95230	(80525, 95321)	x	**RX**	E	DY

NAA PROPELLING CONTROL VEHICLE

Dia. NA508. Mark 1. Class 307 driving trailers converted for use in propelling mail trains out of termini. Fitted with roller shutter doors. Equipment was fitted for communication between cab of PCV and locomotive but this is now isolated and the vehicles are in use as normal vans. B5 bogies. ETH 2X.

Lot No. 30206 Eastleigh 1954–6. Converted at RTC, Derby 1993.

95300	(75114, 94300)	**RX**	E	*E*	EN
95301	(75102, 94301)	**RX**	E	*E*	EN

NOV GENERAL UTILITY VAN (100 m.p.h.)

Dia. NO513. Mark 1. For lot No. refer to original number series. Commonwealth bogies. Add 2 t to weight. ETH0X.

95366	(86251, 93251)	v	**B**	E	Pylle Hill (Bristol)

NRX BAA CONTAINER VAN (100 m.p.h.)

Dia. NR503. Mark 1. Modified for carriage of British Airports Authority containers with roller shutter doors and roller floors and gangways removed. For lot Nos. see original number series. Now used for movement of materials between EWS maintenance depots. Commonwealth bogies. Add 2 t to weight. ETH3.

95400	(80621, 95203)	x	**RX**	E	*E*	CD
95410	(80826, 95213)	x	**E**	E	*E*	EN

NKA HIGH SECURITY GENERAL UTILITY VAN

Dia. NK502. Mark 1. These vehicles are GUVs further modified with new floors, two roller shutter doors per side, middle doors sealed and end doors removed. For lot Nos. see original number series. Commonwealth bogies. Add 2 t to weight. ETH 0X.

95715	(86174, 95115)	**R**	E	*E*	EN
95727	(86323, 95127)	**R**	E	*E*	EN
95734	(86462, 95134)	**RX**	E	*E*	EN
95739	(86172, 95139)	**R**	E	*E*	EN
95743	(86485, 95143)	**RX**	E	*E*	EN
95749	(86265, 95149)	**R**	E	*E*	EN
95754	(86897, 95154)	**R**	E	*E*	EN
95758	(86499, 95158)	**RX**	E	*E*	EN
95759	(86084, 95159)	**RX**	E	*E*	EN
95761	(86205, 95161)	**RX**	E	*E*	EN
95762	(86122, 95162)	**RX**	E	*E*	EN
95763	(86407, 95163)	**R**	E	*E*	EN

NX (GUV) MOTORAIL VAN (100 m.p.h.)

Dia. NX501. Mark 1. For details and lot numbers see original number series. ETH 0 (0X*).

96100	(86734, 93734)	*B5		H		KN
96110	(86738, 93738)	*C		H		KN
96112	(86750, 93750)	*C		H		LT
96130	(86736, 93736)	*C		H		KN
96131	(86737, 93737)	*C		H		KN
96132	(86754, 93754)	*C		H		LT
96133	(86685, 93685)	C		H		LT
96134	(86691, 93691)	C		H		LT
96135	(86755, 93755)	C		H		CP
96136	(86735, 93735)	C		H		LT
96137	(86748, 93748)	C	**B**	H		ZN
96138	(86749, 93749)	C		H		LT
96139	(86751, 93751)	C		H	*VW*	MA
96141	(86753, 93753)	C	**B**	H		LT
96162	(86647, 93647)	*C		H		LT
96163	(86646, 93646)	*C		H		KN
96164	(86880, 93880)	*C		H		LT
96165	(86784, 93784)	*C		H		KN
96166	(86834, 93834)	*C		H		KN
96167	(86756, 93756)	*C		H		KN
96168	(86978, 93978)	*C		H		LT
96170	(86159, 93159)	x*C		H		KN
96171	(86326, 93326)	x*C		H		LT
96172	(86363, 93363)	x*C		H		KN
96173	(86440, 93440)	x*C		H		KN
96174	(86453, 93453)	x*C		H		LT
96175	(86628, 93628)	x*C		H		KN
96176	(86641, 93641)	x*C		H		KN
96178	(86782, 93782)	*C		H		KN
96179	(86910, 93910)	*C		H		LT
96181	(86875, 93875)	*C		H		LT
96182	(86944, 93944)	*C		H		CP
96185	(86083, 93083)	x*C		H		LT
96186	(86087, 93087)	x*C		H		LT
96187	(86168, 93168)	x*C		H		LT
96188	(86320, 93320)	x*C		H		KN
96189	(86447, 93447)	x*C		H		LT
96190	(86448, 93448)	x*C		H		LT
96191	(86665, 93665)	x*C		H		KN
96192	(86669, 93669)	x*C		H		KN
96193	(86874, 93874)	x*C		H		LT
96194	(86949, 93949)	x*C		H		LT
96195	(86958, 93958)	x*C		H		LT

NP (GUV) MOTORAIL VAN (110 m.p.h.)

Dia. NP503. Mark 1. Vehicles modified with concertina end doors. For details and lot numbers see original number series. B5 Bogies. ETH 0X.

96210	(86355, 96159)	H	LT
96212	(86443, 96161)	H	LT
96218	(86286, 96151)	H	LT

AX5G NIGHTSTAR GENERATOR VAN

Dia. AX502. Mark 3A. Generator vans converted from sleeping cars for use on 'Nightstar' services. Designed to operate between two Class 37/6 locomotives. Gangways removed. Two Cummins diesel generator groups providing a 1500 V train supply. Hydraulic parking brake. 61-way ENS interface jumpers. BT10 bogies.

Lot No. 30960 Derby 1981–3. 46.1 t.

96371	(10545, 6371)	**EP**	EU	*EU*	NP
96372	(10564, 6372)	**EP**	EU	*EU*	NP
96373	(10568, 6373)	**EP**	EU	*EU*	NP
96374	(10585, 6374)	**EP**	EU	*EU*	NP
96375	(10587, 6375)	**EP**	EU	*EU*	NP

AY5 (BV) EUROSTAR BARRIER VEHICLE

Dia. AY501. Mark 1. Converted from GUVs. Bodies removed. B4 bogies.

96380–96382/9. Lot No. 30417 Pressed Steel 1958–9. 40 t.
96383. Lot No. 30565 Pressed Steel 1959. 40 t.
96384/6/7. Lot No. 30616 Pressed Steel 1959–60. 40 t.
96385. Lot No. 30343 York 1957. 40 t.
96388. Lot No. 30403 Glasgow 1958–60. 40 t.

96380	(86386, 6380)	**B**	EU	*EU*	NP
96381	(86187, 6381)	**B**	EU	*EU*	NP
96382	(86295, 6382)	**B**	EU	*EU*	NP
96383	(86664, 6383)	**B**	EU	*EU*	NP
96384	(86955, 6384)	**B**	EU	*EU*	NP
96385	(86515, 6385)	**B**	EU	*EU*	NP
96386	(86859, 6386)	**B**	EU	*EU*	NP
96387	(86973, 6387)	**B**	EU	*EU*	NP
96388	(86562, 6388)	**B**	EU	*EU*	NP
96389	(86135, 6389)	**B**	EU	*EU*	NP

NG MOTORAIL LOADING WAGON

Dia. NG 503. These vehicles have been converted and renumbered from weltrol wagons and were used for loading purposes.

Built Swindon 1960. Wagon Lot No. 3102 (3192*).

96452	(B900917)	H	LT

96453	(B900926)	*	H	LT

NVA MOTORAIL VAN (100 m.p.h.)

Dia. NV502. Mark 1. Built 1998–9 by Marcroft Engineering using underframe and running gear from Motorail GUVs. Side loading with one end sealed. The vehicles run in pairs and access is available to the adjacent vehicle. For details and lot numbers see original number series. B5 bogies. ETH 0X.

* Prototype vehicle. Dia NV501. End doors. Now out of use.

99601	(86741, 96101)	*	**H**	H		PY
99602	(86097, 96150)		**GL**	H	*GW*	PZ
99603	(86334, 96155)		**GL**	H	*GW*	PZ
99604	(86337, 96156)		**GL**	H	*GW*	PZ
99605	(86344, 96157)		**GL**	H	*GW*	PZ
99606	(86324, 96213)		**GL**	H	*GW*	PZ
96607	(86351, 96215)		**GL**	H	*GW*	PZ
99608	(86385, 96216)		**GL**	H	*GW*	PZ
99609	(86327, 96217)		**GL**	H	*GW*	PZ

NY EXHIBITION VAN

Various interiors. Converted from various vehicle types. Electric heating from shore supply. In some cases new lot numbers were issued for conversions.

Lot 30842 Swindon 1972–3. Dia. NY503. Converted from BSK to Lot No. 30156 Wolverton 1955.

Non-Standard Livery: Varies according to job being undertaken.

Mk4 denotes a Southern Region Mark 4 EMU trailer bogie.

99621	(34697)	x BR1	**0**	E	OM	Exhibition Coach.
99625	(34693)	x Mk4	**0**	E	OM	Generator Van.

Converted Salisbury 1981 from RB to Lot No. 30636 Pressed Steel 1962. Dia NY523/4 respectively.

99645	(1765)	v C	**0**	E	FK	Club Car.
99646	(1766)	v C	**0**	E	FK	Club Car.

Converted Railway Age, Crewe 1996 from TSO to Lot No. 30822 Derby 1971.

99662	(5689)	B4	**0**	CN	FK

Converted Railway Age, Crewe 1996 from SO to Lot No. 30821 Derby 1971. Originally FO.

99663	(3194, 6223)	B4	**0**	CN	FK
99664	(3189, 6231)	B4	**0**	CN	FK

Converted Railway Age, Crewe 1996 from TSO to Lot No. 30837 Derby 1972.

99665	(5755)	B4	**0**	CN	FK

Converted Railway Age, Crewe 1996 from FO to Lot No. 30843 Derby 1972–3.

99666	(3250)	B4	**0**	CN	FK

YR FERRY VAN

Dia. YR025. This vehicle was built to a wagon lot although the design closely resembles that of NJ except it only has two sets of doors per side. Short Frames (57'). Load 14 t. Commonwealth bogies.

Built Eastleigh 1958. Wagon Lot. No. 2849. 30 t.

889202 **PC** VS *ON* SL

Name: 889202 is branded 'BAGGAGE CAR No.8'.

QSA EMU TRANSLATOR VEHICLES

These vehicles are numbered in the former BR departmental number series but are included here as they are owned by leasing companies and used by them for moving their vehicles around the national system in the same way as other vehicles included in this book. Various diagrams. Converted from Mark 1 TSOs, RUOs and BSKs.

975864. Lot No. 30054 Eastleigh 1951–4. BR Mark 1 bogies.
975867. Lot No. 30014 York 1950–1. BR Mark 1 bogies.
975875. Lot No. 30143 Charles Roberts 1954–5. BR Mark 1 bogies.
975871–975978. Lot No. 30647 Wolverton 1959–61. Commonwealth bogies.
977087. Lot No. 30229 Metro-Cammell 1955–57. Commonwealth bogies.

975864	(3849)		H	*H*	IL
975867	(1006)	**N**	H	*H*	IL
975878	(34643)		H	*H*	IL
975971	(1054)	**P**	P	*P*	CJ
975972	(1039)	**P**	P	*P*	CJ
975973	(1021)	**P**	P	*P*	CJ
975974	(1030)	**N**	A	*A*	IL
975975	(1042)	**P**	P	*P*	CJ
975976	(1033)		A		KN
975977	(1023)		A		KN
975978	(1025)	**N**	A	*A*	IL
977087	(34971)		H	*H*	IL

Mark 4 Stock. Great North Eastern Railway liveried open standard (end) No. 12212 at Doncaster on 24th July 1999 as part of the 09.30 London King's Cross–Edinburgh. **Peter Fox**

▲ **Non-Passenger Carrying Coaching Stock.** Post office sorting van No. 80339 'Brian Quinn' stabled at Carlisle station on 8th June 1999. The vehicle carries Royal Mail livery.
Peter Fox

▼ Post office stowage van No. 80427 stabled at Carlisle on 31st July 1999.
Kevin Conkey

▲ Virgin Trains liveried Mark 3B driving van trailer No. 82145 at Watford at the head of the 10.30 Manchester Piccadilly–London Euston on 1st May 1999.
Kevin Conkey

▼ Mark 4 driving van trailer No. 82206 arrives at London King's Cross on 25th May 1999 at the head of the 12.00 from Edinburgh. **Dave McAlone**

▲ Propelling control van No. 94312 stabled at Carlisle on 8th June 1999.
Peter Fox

▼ High security brake van No. 94403 at Old Oak Common, London on 21st August 1999.
Kevin Conkey

Motorail van No. 96607 on show at London Paddington. This is one of a batch built by Marcroft Engineering for First Great Western services. **Colin J. Marsden**

Nightstar Stock. Nightstar coaches stored at the Alstom Birmingham works (formerly Metro-Cammell) on 15th July 1997. Pictured are seating coach No. 61 19 20-90 029-1 and sleeping car No. 61 19 70-90 028-0. These vehicles are now without work following the decision not to operate Nightstar services. **Peter Fox**

Saloons. Great Western Railway first class saloon No. 9004 is also owned by Railfilms-owned and is seen at Reading on 17th June 1999 in the same train as the photograph on the previous page. **Darren Ford**

Railfilms-owned imitation LMS Club Car No. 99993 (rebuilt from Mark 1 TSO No. 5067) at Reading on 17th June 1999.
Darren Ford

4. NIGHTSTAR STOCK

These coaches were designed for use on new 'Nightstar' services between Britain and Continental Europe via the Channel Tunnel. This new generation of overnight trains was to offer high quality accommodation to both business and leisure customers.

The venture was being developed by European Night Services Limited (ENS), a joint company of Eurostar (UK) Ltd., SNCF, DB and NS. It was originally intended that the trains would operate on the following routes:

London Waterloo–Amsterdam CS.
London Waterloo–Dortmund Hbf./Frankfurt Hbf.
Glasgow/Manchester–Paris Nord.
Plymouth/Swansea–Paris Nord.

Unfortunately the project has now been completely cancelled because of both technical and commercial problems, e.g. there is no locomotive in Belgium which has enough power for the train heating and air conditioning for the length of train envisaged!

Both sleeping cars and reclining seat coaches have been built. Each train was due to be formed of two half-sets, London services having two reclining seat coaches, a service vehicle and five sleeping cars in each half-set to form a sixteen coach train, whilst services from the Provinces to Paris were to be fourteen coaches long with each portion consisting of three sleeping cars, a service vehicle and three reclining seat coaches. The "regional" half-sets were to be numbered 1–9, whilst the "London" half-sets were to be numbered 10–18.

In the following lists, the UIC number for each vehicle is followed by the set number to which it was to belong All stock is the property of Alstom, the manufacturer. Stock is stored at MoD Kineton, MoD Bicester or the Alstom works at Washwood Heath, Birmingham. Current locations are shown where known.

RECLINING SEAT CARS SO End

Each car has 50 seats which are fully reclining, with generous leg space. A table and footrests are provided at each seat. The seats are mounted on plinths and main luggage is stored beneath the seat, while hand baggage is stored in overhead lockers. Individually controlled reading lights are provided, with different levels of ambient lighting for sleeping and non-sleeping hours. Each car has three toilet compartments with shaver sockets .

61 19 20-90 001-0	*1*	KN
61 19 20-90 002-8	*2*	KN
61 19 20-90 003-6	*3*	KN
61 19 20-90 004-4	*4*	KN
61 19 20-90 005-1	*5*	KN
61 19 20-90 006-9	*6*	KN
61 19 20-90 007-7	*7*	
61 19 20-90 008-5	*8*	
61 19 20-90 009-3	*9*	
61 19 20-90 010-1	*10*	
61 19 20-90 011-9	*11*	
61 19 20-90 012-7	*12*	
61 19 20-90 013-5	*13*	
61 19 20-90 014-3	*14*	
61 19 20-90 015-0	*15*	
61 19 20-90 016-8	*16*	
61 19 20-90 017-6	*17*	
61 19 20-90 018-4	*18*	

RECLINING SEAT CAR — SO

Details as above, but no coupling for locomotive.

61 19 20-90 019-2	*1*	KN
61 19 20-90 020-0	*1*	KN
61 19 20-90 021-8	*2*	KN
61 19 20-90 022-6	*2*	KN
61 19 20-90 023-4	*3*	KN
61 19 20-90 024-2	*3*	KN
61 19 20-90 025-9	*4*	KN
61 19 20-90 026-7	*4*	KN
61 19 20-90 027-5	*5*	KN
61 19 20-90 028-3	*5*	KN
61 19 20-90 029-1	*6*	ZE
61 19 20-90 030-9	*6*	KN
61 19 20-90 031-7	*7*	
61 19 20-90 032-5	*7*	
61 19 20-90 033-3	*8*	
61 19 20-90 034-1	*8*	
61 19 20-90 035-8	*9*	
61 19 20-90 036-6	*9*	
61 19 20-90 037-4	*10*	
61 19 20-90 038-2	*11*	
61 19 20-90 039-0	*12*	
61 19 20-90 040-8	*13*	
61 19 20-90 041-6	*14*	
61 19 20-90 042-4	*15*	
61 19 20-90 043-2	*16*	
61 19 20-90 044-0	*17*	
61 19 20-90 045-7	*18*	
61 19 20-90 046-5	*S*	KN
61 19 20-90 047-3	*S*	KN

SLEEPING CARS — SLF End

Each sleeping car has 10 cabins. Six of these have a compact en-suite shower room, with a washbasin, toilet and hairdryers. The remaining four cabins include en-suite toilet and washing facilities, but without the shower.

All cabins are convertible so that when the bunks are folded away by the attendant after passengers have got up, two comfortable armchairs with fold-out tables are revealed. The bunks themselves are generously sized one above the other and will already be made up with duvets, sheets and pillows When passengers arrive. Each cabin has a fitted wardrobe and cupboard, together with facilities for making hot drinks. Cabin telephones are provided for room service.

61 19 70-90 001-9	*1*	KN
61 19 70-90 002-7	*2*	KN
61 19 70-90 003-5	*3*	KN
61 19 70-90 004-3	*4*	KN
61 19 70-90 005-0	*5*	KN
61 19 70-90 006-8	*6*	ZE
61 19 70-90 007-6	*7*	
61 19 70-90 008-4	*8*	
61 19 70-90 009-2	*9*	
61 19 70-90 010-0	*10*	
61 19 70-90 011-8	*11*	
61 19 70-90 012-6	*12*	
61 19 70-90 013-4	*13*	
61 19 70-90 014-2	*14*	
61 19 70-90 015-9	*15*	
61 19 70-90 016-7	*16*	
61 19 70-90 017-5	*17*	
61 19 70-90 018-3	*18*	

SLEEPING CARS — SLF

Details as above, but no coupling for locomotive.

61 19 70-90 019-1	*1*	KN
61 19 70-90 020-9	*1*	KN
61 19 70-90 021-7	*2*	KN
61 19 70-90 022-5	*2*	KN
61 19 70-90 023-3	*3*	KN
61 19 70-90 024-1	*3*	KN
61 19 70-90 025-8	*4*	KN
61 19 70-90 026-6	*4*	KN
61 19 70-90 027-4	*5*	KN
61 19 70-90 028-2	*5*	KN
61 19 70-90 029-0	*6*	ZE
61 19 70-90 030-8	*6*	KN
61 19 70-90 031-6	*7*	
61 19 70-90 032-4	*7*	
61 19 70-90 033-2	*8*	
61 19 70-90 034-0	*8*	
61 19 70-90 035-7	*9*	
61 19 70-90 036-5	*9*	
61 19 70-90 037-3	*10*	KN
61 19 70-90 038-1	*10*	
61 19 70-90 039-9	*10*	
61 19 70-90 040-7	*10*	
61 19 70-90 041-5	*11*	
61 19 70-90 042-3	*11*	
61 19 70-90 043-1	*11*	
61 19 70-90 044-9	*11*	
61 19 70-90 045-6	*11*	
61 19 70-90 046-4	*12*	
61 19 70-90 047-2	*12*	
61 19 70-90 048-0	*12*	
61 19 70-90 049-8	*13*	
61 19 70-90 050-6	*13*	
61 19 70-90 051-4	*13*	
61 19 70-90 052-2	*13*	
61 19 70-90 053-0	*14*	
61 19 70-90 054-8	*14*	
61 19 70-90 055-5	*14*	
61 19 70-90 056-3	*14*	
61 19 70-90 057-1	*15*	
61 19 70-90 058-9	*15*	
61 19 70-90 059-7	*15*	
61 19 70-90 060-5	*15*	
61 19 70-90 061-3	*16*	
61 19 70-90 062-1	*16*	
61 19 70-90 063-9	*16*	
61 19 70-90 064-7	*16*	
61 19 70-90 065-4	*17*	
61 19 70-90 066-2	*17*	
61 19 70-90 067-0	*17*	
61 19 70-90 068-8	*17*	
61 19 70-90 069-6	*18*	
61 19 70-90 070-4	*18*	
61 19 70-90 071-2	*18*	
61 19 70-90 072-0	*18*	

SERVICE VEHICLE/LOUNGE CAR SV

Lounge cars are positioned in each half of the train, between the sleeping cars and the seated accommodation. These vehicles consist of of a sleeping cabin for a disabled passenger and companion with en-suite washroom, a parcels room, offices for train manager and control authority, a lounge with bar for sleeping car passengers and public telephone and a bar for seated passengers.

The vehicle also acts as a base for the sleeping car attendants and for the trolley service which is provided for the seated passengers in the evening. There is also a seated passengers' counter so that snacks and drinks can be obtained during sleeping hours.

61 19 89-90 001-8	*1*	KN
61 19 89-90 002-6	*2*	KN
61 19 89-90 003-4	*3*	KN
61 19 89-90 004-2	*4*	KN
61 19 89-90 005-9	*5*	KN
61 19 89-90 006-7	*6*	KN
61 19 89-90 007-5	*7*	
61 19 89-90 008-3	*8*	
61 19 89-90 009-1	*9*	
61 19 89-90 010-9	*10*	
61 19 89-90 011-7	*11*	
61 19 89-90 012-5	*12*	
61 19 89-90 013-3	*13*	
61 19 89-90 014-1	*14*	
61 19 89-90 015-8	*15*	
61 19 89-90 016-6	*16*	
61 19 89-90 017-4	*17*	
61 19 89-90 018-2	*18*	
61 19 89-90 019-0	*S*	
61 19 89-90 020-8	*S*	

5. SALOONS

Several specialist passenger carrying vehicles, normally referred to as saloons are permitted to run on the Railtrack system. Many of these are to pre-nationalisation designs.

LNER GENERAL MANAGERS SALOON

Built 1945 by LNER, York. Gangwayed at one end with a verandah at the other. The interior has a dining saloon seating twelve, kitchen, toilet, office and nine seat lounge. 21/– 1T. B4 bogies. ETH3.

1999 (902260) **M** GS *ON* EN

GNR FIRST CLASS SALOON

Built 1912 by GNR, Doncaster. Contains entrance vestibule, lavatory, two seperate saloons and luggage space. Gresley bogies. 19/– 1T. 75 m.p.h.

Non-Standard Livery: Teak.

4807 (807) x **0** SH *ON* CJ

LNWR DINING SALOON

Built 1890 by LNWR, Wolverton. Mounted on the underframe of LMS GUV 37908 in the 1980s. Contains kitchen and dining area seating 10 at two tables. Gresley bogies. 75 m.p.h. 10/–.

Non-Standard Livery: London & North Western Railway.

5159 (159) x **0** SH *ON* CJ

GENERAL MANAGER'S SALOON

Dia. AZ501. Renumbered 1989 from London Midland Region departmental series. Formerly the LMR General Manager's saloon. Rebuilt from LMS period 1 BFK M 5033 M to dia. 1654 and mounted on the underframe of BR suburban BS M 43232. B4 bogies. This vehicle has a maximum speed of 100 m.p.h. Screw couplings have been removed. ETH2.

LMS Lot No. 326 Derby 1927. 27.5 t.

Non-Standard Livery: Aircraft blue with gold lining.

6320 (5033, DM 395707) x **0** RV *ON* CP

GWR FIRST CLASS SALOON

Built 1930 by GWR, Swindon. Contains saloons at either end with body end observation windows, staff compartment, central kitchen and pantry/bar. Numbered DE321011 when in departmental service with British Railways. 20/– 1T. GWR bogies. 75 m.p.h.

GWR Lot No. 1431 1930.

9004	x	**CH**	RA	*ON*	CP

WCJS OBSERVATION SALOON

Built 1892 by LNWR, Wolverton. Originally dining saloon mounted on six-wheel bogies. Rebuilt with new underframe with four-wheel bogies in 1927. Rebuilt 1960 as observation saloon with DMU end. Gangwayed at other end. The interior has a saloon, kitchen, guards vestibule and observation lounge. Gresley bogies. 19/– 1T. 28.5 t. 75 m.p.h.

Non-Standard Livery: London & North Western Railway.

45018 (484, 15555)	x	**0**	SH	*ON*	CJ

LMS INSPECTION SALOONS

Built as engineers inspection saloons. Non-gangwayed. Observation windows at each end. The interior layout consists of two saloons interspersed by a central lavatory/kitchen/guards section. BR Mark 1 bogies.

45020–45026. Lot No. LMS 1356 Wolverton 1944.
45029. Lot No. LMS 1327 Wolverton 1942.
999503–999504. Lot No. BR Wagon Lot. 3093 Wolverton 1957.

45026 & 999503 are currently hired to Racal-BRT who have contracted maintenance to the Severn Valley Railway.

45020		**RR**	E	*ON*	ML
45026	v	**M**	E	*ON*	KR
45029	v	**E**	E	*ON*	ML
999503	v	**M**	E	*ON*	KR
999504	v	**E**	E	*ON*	TO

ROYAL SCOTSMAN SALOONS

Mark 3A. Converted from SLEP at Carnforth Railway Restoration and Engineering Services in 1997. BT10 bogies. Attendant's and adjacent two sleeping compartments converted to generator room containing a 160 kW Volvo unit. In 99968 four sleeping compartments remain for staff use with another converted for use as a staff shower and toilet. The remaining five sleeping compartments have been replaced by two passenger cabins. In 99969 seven sleeping compartments remain for staff use. A further sleeping compartment, along with one toilet, have been converted to store rooms. The other two sleeping compartments have been combined to form a crew mess.

Lot. No. 30960 Derby 1981–3.

99968 (10541)	**M**	GS	*ON*	EN	STATE CAR 5
99969 (10556)	**M**	GS	*ON*	EN	SERVICE CAR

RAILFILMS 'LMS CLUB CAR

Converted from BR Mark 1 TSO at Carnforth Railway Restoration and Engineering Services in 1994. Contains kitchenette, pantry, coupé, lounge/reception area with two setees and two dining saloons. 24/– 1T. Commonwealth bogies. ETH 4.

Lot. No. 30724 York 1963. 37 t.

99993 (5067) **M** RA *ON* CP LMS CLUB CAR

BR INSPECTION SALOON

Mark 1. Short frames. Non-gangwayed. Observation windows at each end. The interior layout consists of two saloons interspersed by a central lavatory/kitchen/guards/luggage section. BR Mark 1 bogies.

Lot No. BR Wagon Lot. 3379 Swindon 1960.

999509 **E** E *ON* ML

6. PULLMAN CAR COMPANY SERIES

Pullman cars have never generally been numbered as such, although many have carried numbers, instead they have carried titles. However, a scheme of schedule numbers exists which generally lists cars in chronological order. In this section those numbers are shown followed by the cars title. Cars described as 'kitchen' contain a kitchen in addition to passenger accomodation and have gas cooking unless otherwise stated. Cars described as 'parlour' consist entirely of passenger accomodation. cars described as 'brake' contain a compartment for the use of the guard and a luggage compartment in addition to passenger accommodation.

PULLMAN PARLOUR FIRST

Built 1927 by Midland Carriage and Wagon Company. Gresley bogies. 26/–. ETH 2.

213 MINERVA **PC** VS *ON* SL

PULLMAN BRAKE THIRD

Built 1928 by Metropolitan Carriage and Wagon Company. Gresley bogies. –/30.

232 CAR No. 79 v **PC** NY *ON* NY

PULLMAN PARLOUR FIRST

Built 1928 by Metropolitan Carriage and Wagon Company. Gresley bogies. 24/–. ETH 4.

239 AGATHA **PC** VS SL
243 LUCILLE **PC** VS *ON* SL

PULLMAN KITCHEN FIRST

Built 1925 by BRCW. Rebuilt by Midland Carriage & Wagon Company in 1928. Gresley bogies. 20/–. ETH 4.

245 IBIS **PC** VS *ON* SL

PULLMAN PARLOUR FIRST

Built 1928 by Metropolitan Carriage and Wagon Company. Gresley bogies. 24/–. ETH 4.

254 ZENA **PC** VS *ON* SL

PULLMAN KITCHEN FIRST

Built 1928 by Metropolitan Carriage and Wagon Company. Gresley bogies. 20/–. ETH 4.

255 IONE **PC** VS *ON* SL

PULLMAN PARLOUR THIRD

Built 1931 by Birmingham Railway Carriage and Wagon Company. Gresley bogies. –/42.

261 CAR No. 83 **PC** VS SL

PULLMAN KITCHEN COMPOSITE

Built 1932 by Metropolitan Carriage and Wagon Company. Originally included in 6-Pul EMU. Electric cooking. EMU bogies. 12/16.

264 RUTH **PC** VS SL

PULLMAN KITCHEN FIRST

Built 1932 by Metopolitan Carriage and Wagon Company. Originally included in 'Brighton Belle' EMUs but now used as hauled stock. Electric cooking. B5 (SR) bogies (§ EMU bogies). 20/–. ETH 2.

280 AUDREY **PC** VS *ON* SL
281 GWEN **PC** VS *ON* SL
283 MONA § **PC** VS SL
284 VERA **PC** VS *ON* SL

PULLMAN PARLOUR THIRD

Built 1932 by Metropolitan Carriage and Wagon Company. Originally included in 'Brighton Belle' EMUs. EMU bogies. –/56.

285 CAR No. 85 **PC** VS SL
286 CAR No. 86 **PC** VS SL

PULLMAN BRAKE THIRD

Built 1932 by Metropolitan Carriage and Wagon Company. Originally driving motor cars in 'Brighton Belle' EMUs. Traction and control equipment removed for use as hauled stock. EMU bogies. –/48.

288 CAR No. 88 **PC** VS SL
292 CAR No. 92 **PC** VS SL
293 CAR No. 93 **PC** VS SL

PULLMAN PARLOUR FIRST

Built 1951 by Birmingham Railway Carriage and Wagon Company. Gresley bogies. 32/–. ETH 3.

301 PERSEUS **PC** VS *ON* SL

Built 1952 by Pullman Car Company, Preston Park using underframe and bogies from 176 RAINBOW, the body of which had been destroyed by fire. Gresley bogies. 26/–. ETH 4.

302 PHOENIX **PC** VS *ON* SL

PULLMAN KITCHEN FIRST

Built 1951 by Birmingham Railway Carriage & Wagon Company. Gresley bogies. 22/–.

307 CARINA **PC** VS SL

PULLMAN PARLOUR FIRST

Built 1951 by Birmingham Railway Carriage & Wagon Company. Gresley bogies. 32/–. ETH 3.

308 CYGNUS **PC** VS *ON* SL

PULLMAN FIRST BAR

Built 1951 by Birmingham Railway Carriage & Wagon Company. Rebuilt 1999 by Blake Fabrications, Edinburgh with original timber-framed body replaced by a new fabricated steel body. Contains kitchen, bar, dining saloon and coupé. Electric cooking. Gresley bogies. 14/– 1T. ETH 3.

310 PEGASUS **PC** RA CP

Also carries "THE TRIANON BAR" branding.

PULLMAN KITCHEN FIRST

Built by Metro-Cammell 1960/1 for East Coast Main-line services. Commonwealth bogies. 20/– 2T. 40 t. Those vehicles used in the 'Royal Scotsman' charter train set have been modified and do not carry their names. Their current title and use is shown.

311 EAGLE x **PC** NR *ON* NY On loan to North Yorkshire Moors Rly.
313 FINCH x **M** GS *ON* EN STATE CAR 4 Sleeping Car
317 RAVEN x **M** GS *ON* EN DINING CAR 1 Kitchen & Dining Car
318 ROBIN x **PC** NY *ON* NY
319 SNIPE x **M** GS *ON* EN OBSERVATION CAR Observation Car

PULLMAN PARLOUR FIRST

Built by Metro-Cammell 1960/1 for East Coast Main-line services. Commonwealth bogies. 29/– 2T. 38.5 t. Those vehicles used in the 'Royal Scotsman' charter train set have been modified and do not carry their names. Their current title and use is shown.

324	AMBER	x	**M**	GS	*ON*	EN	STATE CAR 1	Sleeping Car
325	AMETHYST	x	**PC**	FS		SZ		
328	OPAL	x	**PC**	NY	*ON*	NY		
329	PEARL	x	**M**	GS	*ON*	EN	STATE CAR 2	Sleeping Car
331	TOPAZ	x	**M**	GS	*ON*	EN	STATE CAR 3	Sleeping Car

PULLMAN KITCHEN SECOND

Built by Metro-Cammell 1960/1 for East Coast Main-line services. Commonwealth bogies. –/30 1T. 40 t.

335	AMETHYST	x	**PC**	FS	SZ

PULLMAN PARLOUR SECOND

Built by Metro-Cammell 1960/1 for East Coast Main-line services. Commonwealth bogies. –/42 2T. 38.5 t.

347	CAR No. 347	x	**PC**	FS	SZ
348	CAR No. 348	x	**PC**	FS	SZ
349	CAR No. 349	x	**PC**	FS	On loan to Kent & East Sussex Railway
350	CAR No. 350	x	**PC**	FS	SZ
351	CAR No. 351	x	**PC**	FS	SZ
352	CAR No. 352	x	**PC**	FS	SZ
353	CAR No. 353	x	**PC**	FS	SZ

PULLMAN FIRST BAR

Built by Metro-Cammell 1961 for East Coast Main-line services. Commonwealth bogies. 24/– + bar seating 1T. 38.5 t.

354	THE HADRIAN BAR	x	**PC**	FS	SZ

7. COACHING STOCK AWAITING DISPOSAL

This list contains the last known locations of coaching stock awaiting disposal. The definition of which vehicles are "awaiting disposal" is somewhat vague, but generally speaking these are vehicles of types not now in normal service or vehicles which have been damaged by fire, vandalism or collision.

1644	CS
1649	KN
1650	CS
1652	CS
1655	KM
1663	CS
1666	CS
1670	CS
1673	KN
1684	KM
1688	CS
4858	KM
4860	CS
4932	CS
4997	CS
5038	OM
5476	Neville Hill Up Sidings
5505	CS
5533	Neville Hill Up Sidings
5574	Neville Hill Up Sidings
5585	Neville Hill Up Sidings
5595	Neville Hill Up Sidings
6335	LA
6339	EC
6343	HT
6345	EC
6362	LL
6363	LL
6500	Healey Mills Yard
6501	Healey Mills Yard
6521	Healey Mills Yard
6527	Healey Mills Yard
6900	Cambridge Station Yard
6901	Cambridge Station Yard
7183	Crewe Brook Sidings
9458	ZB
6339	EC
9482	NL
13306	KM
13320	CS
17054	Crewe Brook Sidings
18416	Crewe Brook Sidings
18750	Crewe Brook Sidings
19500	Crewe Brook Sidings
21265	KM
34952	SL
35509	ZH
80735	Perth Holding Sidings
80865	Hornsey Sand Terminal
84197	Worksop
84361	Cambridge Station Yard
84364	Doncaster West Yard
84519	Crewe Coal Sidings
92067	ZB
92198	ZB
93180	Derby South Dock Siding
93234	Hayes & Harlington
93358	Mossend Yard
93446	Crewe South Yard
93457	Cricklewood Rubbish Terminal
93482	Bedford Civil Engineers Sidings
93542	Hayes & Harlington
93579	DY
93723	BY
93930	Crewe South Yard
93952	Willesden Brent Sidings
93979	Willesden Brent Sidings
96250	Oxford Hinksey Yard
96256	Oxford Hinksey Yard
96260	Oxford Hinksey Yard
96265	Oxford Hinksey Yard
99648	Eastleigh Locomotive Holding Sidings

8. 99xxx RANGE NUMBER CONVERSION TABLE

The following table is presented to help readers identify vehicles which may carry numbers in the 99xxx range, the former private owner number series which is no longer in general use.

99xxx	*BR No.*	*99xxx*	*BR No.*	*99xxx*	*BR No.*	*99xxx*	*BR No.*
99035	35322	99326	4954	99672	549	99824	4831
99041	35476	99327	5044	99673	550	99826	13229
99052	45018	99328	5033	99674	551	99827	3096
99053	9004	99329	4931	99675	552	99828	13230
99121	3105	99371	3128	99676	553	99829	4856
99125	3113	99405	35486	99677	586	99830	5028
99127	3117	99530	301	99678	504	99880	5159
99128	3130	99531	302	99679	506	99881	4807
99131	1999	99532	308	99680	17102	99886	35407
99241	35449	99534	245	99710	25767	99887	2127
99304	21256	99535	213	99712	25893	99953	35468
99311	1882	99537	280	99713	26013	99961	324
99312	35463	99538	34991	99716	25808	99962	329
99314	25729	99539	255	99717	25837	99963	331
99315	25955	99540	3069	99718	25862	99964	313
99316	13321	99541	243	99721	25756	99965	319
99317	3766	99542	889202	99722	25806	99966	34525
99318	4912	99543	284	99723	35459	99967	317
99319	14168	99545	80207	99782	17007	99970	232
99321	5299	99546	281	99783	84025	99971	311
99322	5600	99566	3066	99792	17019	99972	318
99323	5704	99568	3068	99821	9227	99973	324
99324	5714	99670	546	99822	1859	99974	328
99325	5727	99671	548	99823	4832	99995	35457

The following table lists support coaches and the locomotives which they normally support at present. These coaches can spend considerable periods of time off the Railtrack network when the locomotives they support are not being used on that network.

17007	35028	35207	VS locos	35463	48151	35470	TM locos
17013	60103	35333	6024	35465	D 172	35479	KR locos
17019	45407	35449	34027	35467	KR locos	35486	KR locos
21236	30828	35457	60532	35468	YM locos	80217	75014

LIVERY CODES

Coaching stock vehicles are in Intercity (light grey (white on DVTs)/red stripe/white stripe/dark grey) livery unless otherwise indicated. The colour of the lower half of the bodyside is stated first.

AR Anglia Railways Train Services
B Plain Blue.
BG Blue & Grey.
CC BR Carmine & Cream ('Blood & Custard').
CH BR/GWR Chocolate & Cream.
DR Direct Rail Services. Blue.
DS Danske Statsbaner (Danish State Railways), Blue with moon and star motifs.
E English Welsh & Scottish Railway.
EP European Passenger Services (two-tone grey).
G Southern Region Green.
GN Great North Eastern Railway (dark blue with an orange bodyside stripe and gold or silver GNER lettering).
GL First Great Western (green with gold decals)
GW First Great Western (green with green and ivory vignettes above and below lower bodyside gold band).
H HSBC Rail (blue & white).
LN LNER Tourist green & cream livery.
M BR Maroon.
MD Merseyrail Departmental (dark grey and yellow with Merseyrail logo).
MM Midland Mainline (grey and green with three orange bodyside stripes and Midland Mainline logo and lettering).
N Network SouthEast (grey/white/red/white/blue/white).
NB Network SouthEast livery with the red stripe repainted blue.
O Other livery (non-standard - refer to text).
P Porterbrook Leasing (purple at one end, white at the other. The livery represents an enlarged portion of the Porterbrook logo).
PC Pullman Car Company (umber and cream).
R Plain Red.
RB Regency Rail Cruises livery (Oxford blue & cream).
RM Red with Yellow stripes above solebar with Royal Mail insignia or 'Royal Mail Travelling Post Office' markings.
RP Royal Train Purple.
RR Regional Railways (grey/light blue/white/dark blue).
RX Rail express systems (Post Office red and with Res blue & black markings).
RY Red with yellow stripes above solebar and BR logo.
SS Scotrail sleepers. Two tone purple with silver stripe.
V Virgin Trains (red with black doors extending into bodysides, three white lower bodysides stripes and small Virgin logo adjacent to doors on passenger coaches or red with black inner ends and large full height Virgin logo on DVTs).
WR Waterman Railways (maroon with cream stripes).
WV Waterman Railways VIP (West Coast Joint Stock lined purple lake).

OWNER AND OPERATION CODES

This book now uses a (generally) logical system of codes instead of the gobbledygook codes of the BR Rolling Stock Library (RSL). We have decided to do this since RSL information is not officially available to the general public these days and a system of coding which is fairly obvious to the reader is preferred. For passenger train operating companies these are generally based on those used by Railtrack in the National Rail Passenger Timetable, but there are a few changes for clarity or to reflect changes since the timetable was printed.

OWNER CODES

14	75014 Locomotive Operators Group
24	6024 Preservation Society
62	Princess Royal Locomotive Trust
A	Angel Train Contracts
A4	A4 Locomotive Society
BM	Birmingham Railway Museum
CN	The Carriage and Traction Company
DR	Direct Rail Services
E	English Welsh & Scottish Railway
EU	Eurostar (UK)
FS	Flying Scotsman Railways
GS	Great Scottish & Western Railway Co.
H	HSBC Rail Ltd. (formerly Forward Trust Rail Ltd.)
IE	Ian Storey Engineering
LW	London & North Western Railway Co.
MH	Mid-Hants Railway
MN	Merchant Navy Locomotive Preservation Society
NE	North Eastern Locomotive Preservation Group
NR	National Railway Museum
NY	North Yorkshire Moors Railway
P	Porterbrook Leasing Co. Ltd.
PE	Princess Elizabeth Locomotive Society
RA	Railfilms Ltd.
RS	Rail Charter Services Ltd.
RT	Railtrack
RV	Riviera Trains Ltd.
SH	Scottish Highland Railway Company
SO	Serco Railtest
SP	Scottish Railway Preservation Society
SS	Sea Containers Rail Services
SV	Severn Valley Railway
VS	Venice Simplon Orient Express Ltd.
WC	West Coast Railway Company
WT	Wessex Trains

PERATION CODES

he two letter operating codes give the use to which the vehicle is at present ut. For vehicles in regular use, this is the code for the train operating comany For other vehicles the actual type of use is shown. If no operating code shown then the vehicle is not at present in use.

	Angel Trains Contracts
E	AEA Technology
R	Anglia Railways
A	Cardiff Railway Company
R	Direct Rail Services
	English Welsh & Scottish Railway
U	Eurostar (UK)
N	Great North Eastern Railway
W	Great Western Trains
	HSBC Rail
M	Midland Mainline
W	North Western Trains
	Porterbrook Leasing Co
D	Deicing or Sandite vehicle
N	Used normally on special or charter passenger services
R	Royal Train
S	Locomotive support coach
T	Test Train
L	Silverlink
O	Serco Railtest
R	ScotRail
W	Virgin West Coast
X	Virgin Cross Country
W	Wales & West Passenger Trains

WORKS CODES

ZA	Railway Technical Centre (Derby)
ZB	RFS (E) Ltd., Doncaster
ZC	Adtranz Crewe Works
ZD	Adtranz Derby, Litchurch Lane Works
ZE	Alstom Birmingham, Washwood Heath Works
ZG	Alstom Eastleigh Works
ZH	Railcare Ltd., Springburn Works (Glasgow)
ZN	Railcare Ltd., Wolverton Works
ZP	Bombardier Prorail, Horbury, West Yorkshire

DEPOT TYPE CODE

CARMD	Carriage Maintenance Depot
CSD	Carriage Servicing Depot
TMD	Traction Maintenance Depot
T&RSMD	Traction and Rolling Stock Maintenance Depot

DEPOT & LOCATION CODES

Code	*Depot*	*Operator*
BD	Birkenhead North T&RSMD	Merseyrail Electrics
BK	Bristol Barton Hill T&RSMD	EWS
BN	Bounds Green T&RSMD (London)	GNER
BQ	Bury (Greater Manchester)	East Lancashire Railway
BT	Bo'Ness Station (West Lothian)	Bo'Ness & Kinneil Railway
BY	Bletchley T&RSMD	Silverlink
BZ	St Blazey (Par) T&RSMD	EWS
CD	Crewe Diesel TMD	EWS
CF	Cardiff Canton T&RSMD	Wales & West
CJ	Clapham Yard CSD (London)	South West Trains
CP	Crewe CARMD	London & North Western Rly. Co.
CQ	Crewe (The Railway Age)	Crewe Heritage Trust
CS	Carnforth Steamtown T&RSMD	West Coast Railway Co. Ltd.
CU	Carlisle Currock*	EWS
DI	Didcot Railway Centre	Grest Western Society
DY	Derby Etches Park T&RSMD	Midland Mainline
EC	Craigentinny T&RSMD (Edinburgh)	GNER
EN	Euston Downside CARMD (London)	EWS
FK	Ferme Park CSD	GNER
IL	Ilford T&RSMD	First Gret Eastern
IS	Inverness CARMD	Scotrail
KM	Carlisle Yard *	EWS
KN	MoD Kineton *	Ministry of Defence
KR	Kidderminster	Severn Valley Railway
LA	Laira T&RSMD (Plymouth)	First Great Western
LT	MoD Longtown *	Ministry of Defence
MA	Manchester Longsight CARMD	Virgin Trains
ML	Motherwell TMD	EWS
NC	Norwich Crown Point T&RSMD	Anglia Railways
NL	Neville Hill IC T&RSMD (Leeds)	Midland Mainline
NP	North Pole International (London)	Eurostar (UK)
NY	Grosmont (North Yorkshire)	North Yorkshire Moors Railway
OM	Old Oak Common CARMD (London)	First Great Western
OO	Old Oak Common HSTMD (London)	First Great Western
OY	Oxley T&RSMD (Wolverhampton)	Virgin Trains
PC	Polmadie T&RSMD (Glasgow)	Virgin Trains
PM	St. Phillips Marsh T&RSMD (Bristol)	First Great Western
PY	MOD Pig's Bay (Shoeburyness)*	Ministry of Defence
PZ	Penzance CARMD	First Great Western
RL	Ropley	Mid-hants Railway
SD	Sellafield T&RSMD	Direct Rail Services
SK	Swanwick Junction (Derbyshire)	Midland Railway Centre
SL	Stewarts Lane T&RSMD (London)	Gatwick Express
SZ	Southall (Greater London)	Flying Scotsman Railways
TE	Thornaby T&RSMD	EWS
TM	Tyseley Museum	Birmingham Railway Museum
TS	Tyseley T&RSMD (Birmingham)	Central Trains
YM	National Railway Museum (York)	Science Museum

* Unofficial code